LABORATORY PRACTICES IN MICROBIOLOGY

LABORATORY PRACTICES IN MICROBIOLOGY

Prof. Dr. OSMAN ERKMEN
Department of Food Engineering, Faculty of Engineering
Gaziantep University, Gaziantep, Turkey

ELSEVIER

ACADEMIC PRESS
An imprint of Elsevier

Academic Press is an imprint of Elsevier
125 London Wall, London EC2Y 5AS, United Kingdom
525 B Street, Suite 1650, San Diego, CA 92101, United States
50 Hampshire Street, 5th Floor, Cambridge, MA 02139, United States
The Boulevard, Langford Lane, Kidlington, Oxford OX5 1GB, United Kingdom

Library of Congress Cataloging-in-Publication Data
A catalog record for this book is available from the Library of Congress

British Library Cataloguing-in-Publication Data
A catalogue record for this book is available from the British Library

ISBN: 978-0-323-91017-0

For information on all Academic Press publications
visit our website at https://www.elsevier.com/books-and-journals

Publisher: Andre Gerhard Wolff
Acquisitions Editor: Linda Versteeg-buschman
Editorial Project Manager: Barbara Makinster
Production Project Manager: Kiruthika Govindaraju
Cover Designer: Greg Harris

Typeset by SPi Global, India

Contents

About the author

Born in 1955 in Konya, Turkey, Osman Erkmen is professor of food microbiology in the Department of Food Engineering under the Gaziantep University (Gaziantep, Turkey) since 2004. He received his BS degree in Biology (1985) and MS degree in Food Microbiology (1987) from the Middle East Technical University (Ankara, Turkey). He did his PhD in General Microbiology from the Department of Microbiology Faculty of Medicine under the Gaziantep University in 1994. He started his career as a research assistant at the Department of Food Engineering in 1985 and later became assistant professor in 1994 and associate professor of Food Microbiology in 1999. Since 2004 he is working as professor in this department. At the Department of Food Engineering, he expanded his research to the use of nonthermal processes and natural antimicrobials in food preservation; in the production of fermented foods; in the microbial production of lycopene, thiamin, alcohol, and citric acid from industrial wastes; and in the microbial inactivation and modeling. He received funding for research from the University of Gaziantep Foundation, the Scientific and Technological Research Council, and the Republic of Turkey State Planning Organization. He has been studying the combined effect of nonthermal processes, natural antimicrobials in the destruction of microbial cells and spores, and its application in food preservation, characteristics of white and red wines production from Gaziantep Grapes. He teaches courses in General Microbiology, Food Microbiology, Food Sanitation and

Food Toxicology. Professor Erkmen has published over 100 research articles, reviews, book chapters, proceeding articles, and popular articles in the fields of Food Microbiology, General Microbiology, Food Toxicology and Food Sanitation with more than 3000 citations. He is the editor of the books "Gıda Mikrobiyolojisi" (Food Microbiology; 5th edn) and "Fermente Ürünler Teknolojisi ve Mikrobiyolojisi" (Fermented Products Technology and Microbiology) in Turkish language and is author of one book: Food Microbiology: Principles into Practice (2 volume). He is the advisor of PhD and MS graduates. He has more than 10 patents, organized more than 10 international scientific symposiums and participated more than 50 international symposiums. He held administrative positions as director of the Institute of Natural and Applied Sciences, director of the Institute of Social Sciences, director of the Technical Sciences Vocational School and Head of the Department of Food Sciences.

Preface

The development of good and safe microbiological techniques is essential and expected in the microbiology course. It is also essential for researchers and industrial workers dealing with microbiology. This book has been designed for students and teachers teaching in general microbiology. Various microbiological techniques are described in this book. The techniques of isolation and cultivation of bacteria and fungi, and the determination of their biochemical activities of microorganisms are used in microbiology laboratory practices, researches, and other field area dealing with microbiology. Microbiological techniques are important in identifying the microorganisms from food, clinical, and environmental samples. In this book, each practice was started with a brief introduction revealing the theoretical basis of the subject, so that there will be a strong conjunction between the practical and theoretical sessions. The comprehensive procedures are given in the practices. The practices have been designed to use commonly available equipment, least expense involved, and be completed in the shortest possible time period. Practices also include laboratory report that help students to record paratactical laboratory results, evaluate results, prepare laboratory report, and gain more information about the practices. It is hoped that the book will serve varied departments dealing with microbiology such as Biology, Food Engineering, Faculty of Health Science, Agricultural Engineering, Food Technology, Environmental Engineering, Nutrition and Dietetic Department, Food Hygiene, as well as anyone interested in different branches of biology, health, agriculture, and food production. The book is divided into five sections with different practices: Section I, basic microbiological techniques with five practices; Section II, morphology of bacteria with eight practices; Section III, biochemical characteristics of microorganisms with three practices; Section IV, chemical and physical effects on microorganisms with seven practices, Section V, fungi and bacteriophages with two practices. Appendix A is of great value, including directions for the preparation of special media, stains, and solutions.

Corrections, technical questions, and interests could be sent to: osmerkmen@gmail.com.

Osman Erkmen

Acknowledgments

I always thank my students who like microbiology lessons because the desire of my students to learn and access information inspired me to write this book. I would like to thank my wonderful wife Asst. Prof. Dr. Ayşe Erkmen and our sons Dr. Barış Erkmen and Electric Electronic Engineer Hüseyin Erkmen for their support, sincere encouragement, and endurance.

Finally, I would like to thank the ubiquitous and invisible microorganisms, I wrote and published a book because of their existence, and also had to join the "real world."

Introduction to the student

Materials to be provided by student

1. A clean laboratory coat,
2. Laboratory book,
3. A water-proof marking pen,
4. A pen
5. A laboratory notebook and
6. Special requirements that may be indicated by laboratory instructor.

General laboratory rules

1. Each laboratory class will begin promptly at the indicated time.
2. If you are allergic to any antibiotics or others, please inform the instructor immediately.
3. Always wear a laboratory coat during working in the laboratory.
4. You are expected to read the assigned laboratory practice and carry out the suggested reading assignments before each class. The instructor will design the parts of a technique scheduled to be undertaken in class. Laboratory instructions will not be repeated if you are late. Do not forget your laboratory book. Wait for a laboratory introduction by the instructor before starting practices.
5. Long hair should be tied back while in laboratory.
6. Do not leave valuables in coat packets. Food must not be eaten in the laboratory. Do not drink from laboratory ware.
7. Never place any materials used in the laboratory, including pencils, pens and markers, in the mouth.
8. All materials and clothes other than those needed for the laboratory are to be kept away from the work area.
9. Smoke will not be tolerated in the laboratory.
10. Microbiological cultures must not be removed from the laboratory.
11. Store your personal belongings, such as textbooks and coat, in the area specified by your instructor. Never let them clutter the operating space on the laboratory bench, where they might become contaminated through exposure to microbial culture.

12. The needle or loop and other equipment used in transferring cultures should be sterilized in the flame before and after use. Flame the wire vertically, not horizontally. If it is covered with viscous material, dry at the side of the flame before burning, to avoid scattering living material.
13. In case a living culture is spilled, immediately notify the instructor in charge. It should be covered with an efficient disinfectant. The hands should then be disinfected and washed in soap and water.
14. All noninfectious solid wastes, such as cotton, paper, and matches, should be placed in the receptacles provided for this purpose.
15. All discarded cultures should be placed in the appropriate containers and sterilized in the autoclave.
16. In case of personal accident, such as cutting or pricking the finger, or splashing culture material in the eye, report immediately to the instructor in charge.
17. At the end of practice, put away all apparatus.
18. Unless your instructor specifies otherwise, each practice report is to be completed within 1 week. Your laboratory book and notebook are mean to serve as a record of your work.
19. Leave all laboratory facilities and equipment in good order at the close of each period. Inform the instructor of any defects in equipment.
20. All accidents, cuts, and any damaged glassware or equipment should be reported to the instructor immediately.
21. At the end of the microscopic practice, carefully clean the oil-immersion lens using only special lens tissue. Return the microscope to its box. Replace reagent bottles in their proper place. All used equipment must go in the appropriate receptacles.
22. Clean the laboratory table before and after laboratory hour with the 10% bleach (hypochlorite) solution provided.
23. Before leaving the laboratory, see that the gas and water are turned off.
24. Wash the hands thoroughly with soap and water, disinfect using disinfectant and dry out before leaving the laboratory.
25. Keep complete records of each day's work, with drawings, in a standard laboratory notebook.
26. Answer the questions indicated at the end of each practice.
27. You must write reports with ink or ball pan.
28. It is the responsibility of the student to know the location and use of all safety equipment in the laboratory (eyewash, fire extinguisher, etc.)

Laboratory safety

1. Carry and/or store microbial cultures and inoculated media in racks, baskets, or other designated containers. Do not lay liquid cultures on table tops. Avoid spillage.
2. Always flame inoculating loops and/or needles after use.
3. Place contaminated materials, old cultures, and the results of exercises into the containers designated by the instructor.
4. In the event of a laboratory accident such as the spilling or dropping of a live culture, remain calm. Carry out the following procedure:
   - **(a)** Report the accident to your instructor as soon as possible.
   - **(b)** Place paper towels over the spilled material.
   - **(c)** Pour disinfectant liberally over the towels.
   - **(d)** After 15 min, remove and dispose of the towels in the receptacle for the disposal of contaminated materials.

SECTION I

Microbiological media and cultural techniques

Most microbiological laboratories prepare their microbiological media. These media will be given to you in the laboratory hour. In some cases, special media will be needed. They will also be given to you. The preparation of media in the laboratory requires sterilization and provision of aseptic conditions. Adding a culture from one media to another also requires aseptic conditions. Aseptic condition is the prevention of contamination of microorganisms from the environment while working microbiologically. The media on which the microorganism will be reproduced can be solid and liquid. The growth of microorganisms in liquid media is detected by the formation of turbidity, and in the solid media, by the formation of colony. Isolating a microorganism away from other microorganisms involves many cultural techniques. The teaching of such an important subject as microbiology cannot be achieved effectively without enhancing the theory with hands-on experience in the laboratory. The purpose of this section is to provide good techniques in practical microbiology to ensure that investigations proceed safely and achieve the required educational aims successfully.

The microbiological practices given in this section are as follows:

Practice 1. Preparation of Media and Sterilization Techniques

Practice 2. Microorganisms in Environment

PRACTICE 1

Preparation of media and sterilization techniques

1.1 Theoretical information

A microbiological medium (plural: media) contains the nutrient and is used for culturing bacteria, yeasts, molds, and algae. Media contain necessary nutrients supporting the growth (replication) of microorganisms. Media are prepared in liquid, semisolid, and solid form. They are used for several purposes: (a) growth (culture) media, (b) diagnostic (identification) media, (c) transport media, (d) storage media, and (e) assay media.

1.1.1 Media for microorganisms

Media can be prepared in three forms based on the physical state: Liquid, solid and semisolid.

Liquid media. They are often prepared from different nutrient components, such as peptone, yeast extract, inorganic salts, distilled water, and others. Nutrient broth, brain heart infusion broth and potato dextrose broth are examples of liquid media that are used for the growth of microorganisms. Bacteria, many algae, and fungi may be cultured in liquid media.

Solid media. They are prepared by adding a solidifying agent (1.5%–2% of agar) such as agar, gelatin, or silica gel, to the liquid media. Solidifying agent, not used by microorganisms and does not inhibit microorganisms, is an acidic polysaccharide and does not liquefy at room temperature. An agar gel is formed when a mixture of agar in the water heated above 80°C and subsequently cooled; solidified at about 45°C. Agar is obtained from certain species of Rhodophyta (algae, such as *Gelidium*). Examples of solid media are nutrient agar, brain heart infusion agar, and potato dextrose agar. Bacteria, yeasts, and molds can grow on the solid media and produce colonies.

Semisolid media. They contain less amount of solidifying agent (0.5% of agar) and present between liquid and solid media is called semisolid (jelly-like) medium. Semisolid media are used for special purposes such as

Laboratory Practices in Microbiology
https://doi.org/10.1016/B978-0-323-91017-0.00004-4

transfer of culture and indication of motility. Fluid thioglycolate medium and motility medium are examples of semisolid media that are used for the identification of certain characteristics of microorganisms.

1.1.2 Nutritional requirements of microorganisms

Before preparing a medium, one must understand the basic needs of microorganisms for growth. In the selection of any suitable medium for a specific group of microorganisms, the following basic requirements must be considered: water, energy, carbon, nitrogen, minerals, growth factors, and pH.

Water. Microbial cell protoplasm consists of 70%–85% of water. The water is essential for a microorganism for the following functions: retaining viability, transport of nutrients, removal of wastes, free movement of molecules in cytoplasm and on membrane, predicting mechanical structure, and metabolic reactions. The quality of water used in the preparation of media is important. Hard tap water (containing a higher amount of calcium and magnesium ions) should not be used. Ions can precipitate in the presence of peptones and beef extract. Therefore, the best way is to always use distilled water in the preparation of media.

Carbon. Microorganisms are divided into two groups depending on the use of carbon. Microorganism utilizing carbon dioxide as a carbon source for synthesis of cellular components is called autotroph. If the microorganism uses organic compounds as carbon source, it is called heterotroph. Specific organic carbon needs are different among microorganisms. As one microorganism may require only a simple organic compound, another requires complex organic compounds.

Energy. Photosynthetic bacteria synthesize organic matter as a result of photosynthesis events using light energy from the sun. Photosynthesis occurs in microorganisms containing chlorophyll pigment. Chemosynthetic bacteria perform chemosynthesis events using chemical energy. They synthesize organic nutrients by oxidation of inorganic substances in the presence of oxygen. They do not contain chlorophyll pigment and not use light energy. Photosynthetic and chemosynthetic bacteria synthesize organic compounds (e.g., glucose) from inorganic substances.

Nutrition of organisms according to the combination of energy and carbon sources is given in Table 1.1. Some microorganisms are called photoautotrophs that use light energy from the sun as an energy source, CO_2 as a carbon source, and H_2O, H_2S, or H_2 as a hydrogen source. Some Protista, some bacteria, and algae are examples of photosynthetic microorganisms. Some microorganisms obtain energy by oxidizing simple inorganic

Table 1.1 Energy and carbon sources in different types of nutrition.

Nutrition type	Energy supply	Carbon source	Example
Photoautotroph	Light	CO_2	Algae, cyanobacteria, plants
Chemoautotroph	Inorganics (H_2S, NH_3, Fe^+, etc.)	CO_2	A few bacteria and most archaea
Photoheterotroph	Light	Organic compounds	Purple non-sulfur bacteria, green non-sulfur bacteria, heliobacteria.
Chemoheterotroph	Organic compounds	Organic compounds	Many bacteria, fungi, some archaea, animals, humans.

substances (e.g., nitrite, nitrate, and sulfur) and use CO_2 as the source of carbon and H_2O as the source of hydrogen, they are called chemoautotroph. Some microorganisms that use sun rays as an energy source and organic compounds as a carbon source are called photoheterotrophs (e.g., certain aquatic and salt-loving prokaryotes, some bacteria, and arches). Some microorganisms that use organic compounds (such as glucose or amino acids) as carbon and energy sources are called chemoheterotrophs (e.g., fungi, most of the Protista, many bacteria, and archaea).

Nitrogen. Autotrophic microorganisms utilize nitrogen; the heterotrophs get their nitrogen from amino acids, proteins, peptides, proteases, peptones, and beef extract.

Minerals. All microorganisms require minerals in very small amounts (e.g., potassium, sodium, magnesium, calcium, manganese, copper, phosphorus, iron, zinc, and cobalt) for growth.

Growth factors. Any essential component required by a microorganism that is unable to synthesize is called growth factor. They are some types of amino acids or vitamins. Many heterotrophs grow in nutrient broth and other media since they contain essential growth factors. Most pathogens grow in an enriched media containing growth factors, such as blood agar.

Hydrogen ion concentration. Some of the microorganisms cannot grow at low pH if the pH of the medium is not within their growth range. The microbial enzymes are greatly affected by hydrogen ion concentrations. Most bacteria grow best around neutral environment (pH7.0 or slightly lower).

1.1.3 Types of media

Microbiological media can be prepared from known exact specifications of ingredients. They are made from highly purified and defined chemical compounds. Both the beef extract and peptone in nutrient broth are exact in composition. They are called synthetic media. Media containing imprecise composition of ingredients are called non-synthetic media.

(1) General purpose media. They usually contain a source of carbon, energy (often glucose), salts, amino acids, and vitamins. Different complex raw materials (such as peptone, meat extract, and yeast extract) are used in the preparation of these media. These media support the growth of different microorganisms. More general-purpose media are used to enumerate and isolate certain types of microorganisms. Nutrient broth/agar, tryptose soy broth/agar, brain heart infusion broth/agar, and Sabouraud dextrose broth/agar are examples of the general-purpose media.

(2) Anaerobic media. Some microorganisms are anaerobes that tolerate little or no oxygen. A reducing agent (such as sodium thioglycolate) is added to the media that would remove oxygen and to make reduced media. Reducing agent chemically combines with oxygen dissolved in the medium and makes it unavailable for microorganisms. During the anaerobic media preparation, they are boiled to drive off most of the dissolved oxygen. Thioglycolate medium, cooked meat medium, tryptic soy anaerobic medium, and brewer anaerobic agar are examples.

(3) Transport media. They are used during the transport of cultures to the laboratory. Transport media ideally maintain the viability of microorganisms without altering their number. They typically contain only buffers and salts for specific microorganism. The lack of carbon, nitrogen, and organic compounds in media prevents the microbial growth. Transport media used for anaerobes must be free of molecular oxygen. Amies media, aerobic transport media, and anaerobic transport media are examples.

(4) Enriched media. They contain nutrients to support the growth of a wide variety of microorganisms. Enrich provides an environment with both chemical and physical agents to increase numbers of microorganisms. No inhibitory agent is used to prevent the growth of unwanted microorganisms. Enriched media (e.g., enriched broth, selenite broth, and Gram-negative broth) hold the normal flora (unwanted microorganisms) in the lag phase while allowing the logarithmic growth of the microorganisms. After enrich, the required microorganisms can be isolated by pure culture techniques.

(5) Selective media. They allow certain types of microorganisms to grow because of (1) the absence of certain critical nutrients in media that make it unfavorable for most of the microorganisms, (2) the inhibitory substances in media prevent the growth of unwanted microorganisms, and (3) providing rapid isolation and identification aids to the microorganisms by inhibiting growth of the unwanted microorganisms. The inhibitory substance added into media may be salt, acid, dyes, antibiotic (e.g., streptomycin, kanamycin, chloramphenicol, and gentamicin) and others. For example, Potato dextrose agar selects fungi because it has a low pH of 5.6 and a high glucose concentration. Crystal violet or brilliant green dye is added to inhibit the growth of Gram-positive bacteria without affecting the growth of Gram-negative bacteria. On the other hand, phenylethanol agar inhibits the growth of Gram-negative bacteria, but not that of Gram-positive bacteria. Antibiotics are added to some media to make them selective for microorganisms that are resistant to these antimicrobial agents, such as Salmonella-Shigella agar and CIN agar.

(6) Differential media. They contain substances that cause formation of different colony appearance of species. These media help differentiate various kinds of microorganisms on same agar plate by their colony appearances. For example, if a mix-culture of bacteria is inoculated onto a blood-agar, some bacteria may produce enzymes that hemolysis the red blood cells, for hemolytic zone around colonies and others may not. Based on the clear (hemolytic) zone formation, one can distinguish between hemolytic and nonhemolytic bacteria. Blood agar and starch agar are examples.

(7) Selective/Differential media. They are both selective and differential. They differentiate microorganisms depending on the selective agent and appearance on/in the media. Eosin-methylene blue agar contains one sugar (lactose) and a dye (eosin-methylene blue) which turns red when the pH drops below 6.8. Any microorganism that can ferment lactose produces an acid that lowers the pH and causes the colony to turn red with metallic-sheen in the presence of dye. Microorganisms that fail to use lactose produce colorless colonies. Mannitol salt agar and MacConkey's agar are other examples.

(8) Assay media. They are used to detect concentrations of substance (such as antibiotics and vitamins). For example, Mueller-Hinton agar and vitamin assay media can be used to determine the susceptibility of microorganisms to antibiotics and vitamin production level, respectively.

1.1.4 Preparation of media

Most of the media used by students are prepared in the microbiology laboratory (Fig. 1.1). The equipment used in the preparation of media are given in Fig. 1.2. In the preparation of media, first, determine how much ingredients are needed for the preparation of 1000 mL of medium. The required amount of ingredients is weighed on a scale and added into the water in a

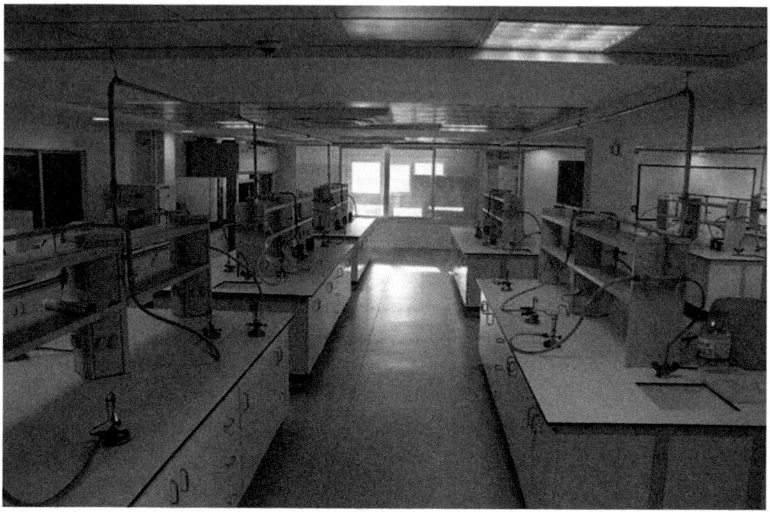

Fig. 1.1 A microbiology laboratory room.

beaker. If the medium does not contain agar, the mixture easily dissolves in water without heating and the pH of medium is adjusted using pH meter and magnetic stirrer. If the medium contains agar, it is prepared in flask. The agar medium is heated to melt agar near boiling (as gas bubbles start to appear; "froths up") over a Bunsen burner or on an electric heater (heating mantle). The level of the top of the medium is marked on the side of the beaker with a waterproof marking pen before heating, to indicate the loss of water. The level is completed with using hot distilled water. As soon as the medium "froths up," the heater is turn off and the medium is removed from the hot plate. Media can be prepared in the form of broth, agar slant, stab agar and Petri plate (petri dish containing agar medium). Petri plate is prepared by pouring sterile agar medium into sterile Petri dish. The followings can be taken into consideration in the preparation of media:

Medium	Amount	Container	Sterilization
Broth	5–10 mL	Tube	by autoclave
Agar slant	8 mL	Tube	by autoclave
Stab agar	7 mL	Tube	by autoclave
Petri plate	18–22 mL	Petri dish	Agar media in autoclave and Petri dishes in dry-air oven

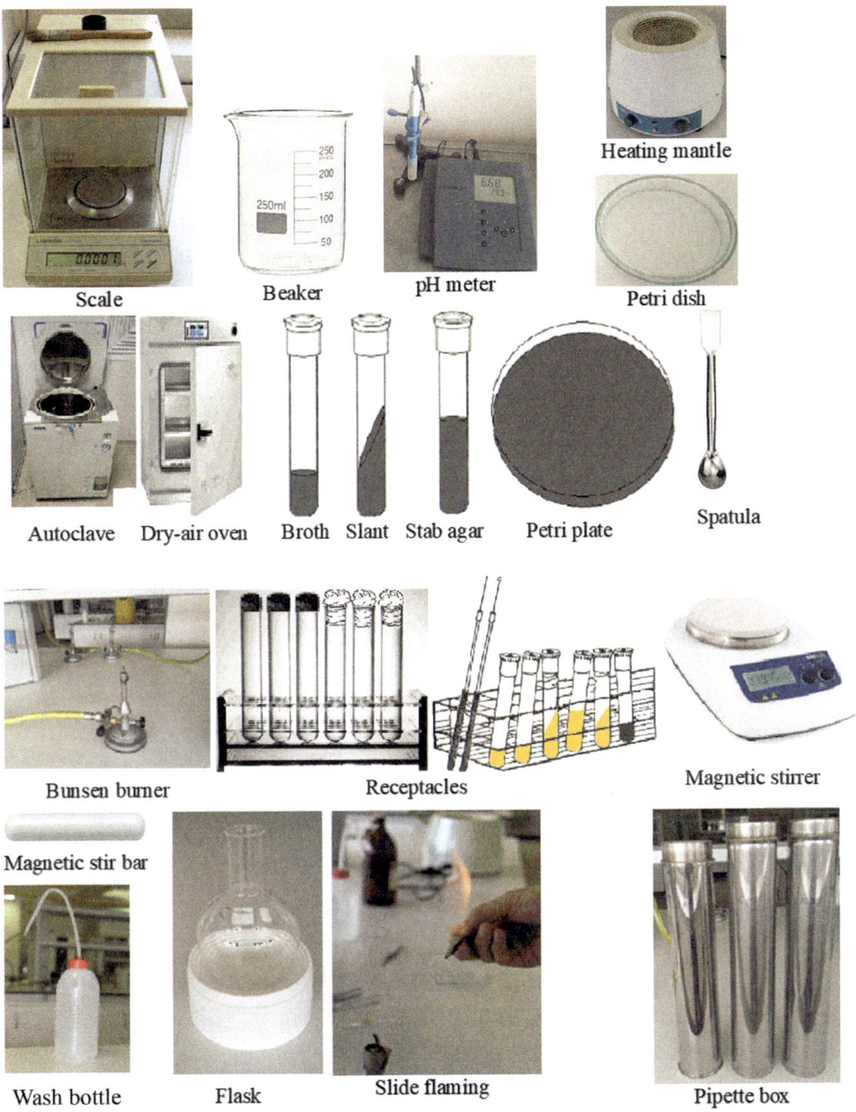

Fig. 1.2 Some types of media and equipment used in the microbiology laboratory.

1.1.5 Sterilization of media and equipment

Sterilization is a process in which all living microorganisms, viable spores, viruses, and others are destroyed (killed). The microorganisms and agents are killed with heat (such as steam, dry heat, and incineration) and chemicals. Microorganisms are abundant in the environment. They are on your body, in air, and on every surface and object that you encounter. The systematic study of isolated microorganisms would be impossible without a method of destroying microorganisms. Without effective sterilization procedures, aseptic techniques and microbiological laboratory studies would not be possible.

1.1.5.1 Physical techniques of sterilization

The most common physical techniques of sterilization are moist heat, dry heat, and filtration.

(a) Heat

Heat is generally preferred as sterilizing agent; it is non-selective on all microorganisms including spores and others. The lethal action of heat is due to the oxidation-burning or coagulation-denaturation of microbial proteins and nucleic acids. The effects of heat on the lipids of a cell membrane may also occur.

(i) Wet sterilization

Moist heat is more lethal than dry heat; it requires low working temperatures and shorter periods of exposure. Moist heat denatures and coagulates microbial proteins (such as enzymes). Denaturation of cellular proteins occurs at lower temperatures and shorter exposure times. The stabilizing protein bonds (e.g., C=O...H–N) are more easily broken when water molecules are available for hydrogen bonding. Moist heat used to kill microorganisms can be in the form of steam, boiling water (at 100°C), and water heated to sub-boiling temperature (pasteurization). Moist heat sterilization involves an efficient transfer of heat since steam which condenses on a cooler object delivers up a considerable quantity of (latent) heat. Moist heat is 2500 times more efficient sterilant than hot dry air at the same temperature.

Wet sterilization is commonly carried out by steam in the autoclave (Fig. 1.2). The autoclave is used to sterilize the usual solid and liquid media, solutions, discarded cultures, contaminated glassware, heat stable rubber tubing, cotton, and others. An autoclave is a large double-jacked pressure cooker. When the autoclave is closed and the heat is turned on, the water heats and pressure build up in the sealed autoclave. As the pressure increases above normal atmospheric level, the boiling point of water is raised. For example, at 15 pounds per square inch (psi) above normal atmospheric pressure at sea level, water boils at 121°C instead of 100°C as it does at atmospheric pressure at sea level. The suitable sterilization in the autoclave is performed by maintaining a temperature of 121°C for 15–20 min at 15 psi (1.02 atm). Some media, such as certain fermentation broth, nutrient gelatin, and litmus milk, deteriorate at 121°C. They must be autoclaved at a lower temperature and pressure (such as at 115°C for 10 min).

(ii) Dry sterilization

Dry heat is generally used in three ways: dry-air, flaming, incineration, and baking. Incineration is the burn-up microorganisms and agents. Incineration is useful for destroying heavily contaminated materials or tissues. Dry heat or hot air used in dry-air oven (Fig. 1.2) kills all microorganisms at high temperatures. It is preferred for sterilizing clean glassware (e.g., Petri dishes, pipettes, flasks, and test tubes) because it cannot cause condensation of moisture on glassware. Pipettes are sterilized in pipette box, empty tube in receptacle and petri dish together with lid (Fig. 1.2). Dry heat causes coagulation and oxidation of organic constituents of microbial cells and then it can burn them slowly. Hence, this approach is not as effective as moist heat, and much higher temperatures and longer times are necessary to burn organic matter. Dry sterilization is carried out in a dry-air oven at 180°C for 2 h.

(iii) Tyndallization

Heat-label solution or medium which would be adversely affected by autoclaving can be sterilized through Tyndallization. In this method, the material is heated to 90–100°C for several minutes on each of three successive days, being incubated at 35°C for intervening periods. The first period of heating kills all vegetative cells initially present in the material, incubation during intervening periods permits germination of any spores which may survive at heating; vegetative cells that arise from such spores are killed during the subsequent heating.

(iv) Naked flame

The naked flame of the Bunsen burner is used for sterilizing the platinum needle and loop (Fig. 1.2). They are heated to redness. Mouth of glassware, test tubes, and others are disinfected before and after use with a very short exposure through flame (known as flaming). Small instruments, glass spreader, microscope slides, cover glasses, forceps, and others are disinfected by dipping them in alcohol (Fig. 1.2) and then burning off the surface of instruments.

(b) Bacteriological filtration

Liquid and gasses can be sterilized by passing them through filters. It is useful method for sterilizing heat labile liquids, enzymes, and solutions. This method is based on mechanical process of removing microorganisms from solution by passing it through a filter ($\leq$ 0.2 or 0.4 μm pore size) and microbial cells are retained on the filter. Cotton-wool filters can also be used for disinfection of air and gasses.

1.1.5.2 Chemicals for sterilization and disinfection

Many chemicals are used as disinfectants (as indicated in Practice 3). Their lethal effect on microorganisms is due to several mechanisms including the denaturation of protein and enzymes; and destruction of DNA, cell walls, and cytoplasmic membranes. Some chemicals can be used for the disinfection of working area in the laboratory. Container of disinfectant (such as zephran solution, hypochlorite solution, and synthetic detergent) is supplied in the laboratory, usually beside the sinks. You should make it a habit to deep your hands into a disinfectant and then wash after you finish practice in laboratory. Examples of disinfectants and their applications:

Disinfectant	Use	Concentration
Virkon	Surfaces, slides and skin disinfection.	1.5% (v/v)
Hypochlorite	In discard pots for pipettes and slides	0.25% v/v) free chlorine
Ethanol	Skin disinfection	70% (v/v)
Zephran	Skin disinfection	0.1% (v/v)

Three chemicals can be recommended as sterilizing agents' ethylene oxide, formaldehyde, and glutaraldehyde. They kill all microorganisms including bacterial endospores and viruses. So, this sterilization is called "cold sterilization". These agents must be used in effective concentrations and under appropriate conditions.

1.2 Techniques in the preparation of media

Materials

Media: Nutrient agar and nutrient broth.

Equipment. Magnet stirring, balance, Scale, distilled water, flask, test tubes, test tube holder, sterile pipettes in box (1 and 10 mL), sterile petri dishes, dry-air oven, autoclave, incubator (at 35°C), pH meter, graduated cylinder, tartaric acid, and spatula.

1.2.1 Nutrient broth preparation

1. Calculate necessary ingredients to make 100 mL of nutrient broth.
2. Weighing ingredients on balance with a clean spatula and put into 250 mL beaker (Fig. 1.2).
3. Add 100 mL water, put magnetic stir bar (Fig. 1.2) into medium in the beaker and place beaker onto magnetic stirrer (Fig. 1.2) and mix ingredients.

Fig. 1.3 Aluminum foil covered Erlenmeyer and receptacle before sterilization.

4. Adjust pH of medium to 6.8 using pH meter (Fig. 1.2) with 0.1 N NaOH and 5% tartaric acid (or 0.1 N HCl).
5. Fill test tubes (bottles) with 5 mL of liquid medium.
6. The mouth of each tube is closed with closer and test tubes are placed on the test tube holder (Fig. 1.2). Serializable plastic (polypropylene) screw-caps (or cotton-plug) are used in closing the tubes. Before sterilization, the plastic screw caps should not be screwed tightly, and each one must be left partly unscrewed and providing an air gap to allow steam to escape during sterilization. If cotton-plug is used in closure of tubes, a properly made cotton plug should hold firmly in the tube so that it is not easily dislodged. Cotton plugged tubes on the receptacle should be covered with aluminum foil (Fig. 1.3) and tied to prevent steam condensation on cotton during steam sterilization.
7. Prior to sterilization, the tube holder should be placed in a wire basket (Fig. 1.4) in the autoclave. Label the media with the type of medium and your name or initials with waterproof marking pen. Sterilize all media in the autoclave.

1.2.2 Nutrient agar preparation

1. Calculate necessary ingredients to make 200 mL of nutrient agar.
2. Weigh ingredients on balance with a clean spatula and put into 250 mL beaker except agar.
3. Add 200 mL water, put magnetic stir bar (Fig. 1.2) into stirrer into beaker and place beaker onto magnetic stirrer (Fig. 1.2) and mix ingredients.
4. Adjust pH of medium to 6.8 with sterile 0.1 N NaOH and 5% tartaric acid (or 0.1 N HCl) and put solution into 250 mL flask (Fig. 1.2) then add agar (1.5%, w/v).

Fig. 1.4 Media sterilization in a wire basket in the autoclave.

5. Bring the solution to a boil by placing the flask or on an electrical heater mantle (Fig. 1.2) and continue to heat until the agar is molten. Make up the loss with boiling distilled water due to evaporation.
6. Place test tubes onto receptacle (Fig. 1.2). Transfer 7 and 8 mL of molten media into tubes.
7. Repeat steps 1–5 for the preparation of nutrient agar in 500 mL flask (Fig. 1.2).
8. Close the neck of tubes by screw-caps or cotton-plug and flasks by cotton-plugs indicated in nutrient broth preparation as in step 6.

Prior to sterilization, the tube receptacle and flask should be placed in a wire basket (Fig. 1.4) in the autoclave. Label the media with the type of medium. Sterilize all media in the autoclave.

1.2.3 Sterilization

(i) Sterilize all media and solutions in the autoclave at 121°C for 15–20 min (Fig. 1.4) and (ii) sterilize all glassware in the dry-air oven at 180°C for 2 h (Fig. 1.2).

1.2.4 After sterilization

1. *Nutrient agar plates.* Pour exactly 20 mL of nutrient agar (cooled to 48°C) from flask into sterile Petri dishes (90 × 17 mm) under aseptic conditions near Bunsen burner flame (Fig. 1.5) and allow to solidify the agar; then

invert the plate to prevent condensation of vapor from lid. Label the plate with the name of medium.

2. *Nutrient agar in tubes*. After sterilization, plastic caps of tubes are screwed tightly. If tubes of agar media (7 mL) are to be converted to slants, it is necessary to lay the tubes down in a near-horizontal manner (at a slanted position) as soon as they are removed from the autoclave. The easiest way to do this is to use a piece of rubber tubing (1.25 cm) to support the capped end of the tube (Fig. 1.6) and allow it to solidify the agar. Remaining agar in tubes (8 mL) allowed upright to cool and used as nutrient stab agar (Fig. 1.2). After cooling, label the tubes with name of medium and place test tubes to tube holder.

Fig. 1.5 Pouring about 20 mL sterile agar medium (at 48°C) into sterile Petri dishes under aseptic conditions.

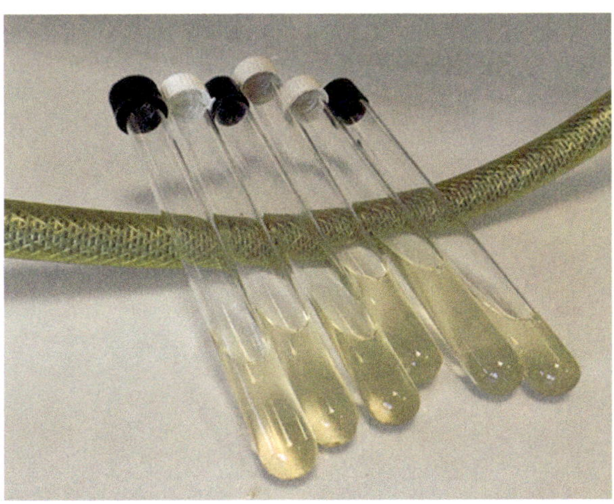

Fig. 1.6 Nutrient agar slant preparation.

Fig. 1.7 Incubation of media in the incubator.

3. *Nutrient broth in tubes.* After sterilization, the plastic caps of tubes are screwed tightly. Tubes of broth should be allowed to cool at room temperature (Fig. 1.2). After cooling, label the tubes with name of medium, and place test tubes onto receptacle.
4. Once prepared media have cooled, place test tube holder and inverted Petri plates into an incubator (Fig. 1.7) at 35°C for overnight to control possible contamination.
5. If media in tubes and plates are not used immediately, they should be stored in a refrigerator. The plates are inverted during incubation and storage to prevent condensation of vapor on lid and drop of water onto medium. Water condensate onto medium during incubation may contaminate microorganisms or spread colonies on the surface area of the medium. When they are stored for long periods at room temperature, media tend to lose moisture.

LABORATORY REPORT

Date:
Name:

Lab Title: Preparation of Media and Sterilization Techniques

1. **Record your results about the sterility and contaminants of media. Which signs would indicate contamination in the agar medium if present after incubation?**

2. **Explain why it is useful to study microorganisms in the laboratory and not in natural habitats.**

3. **Define the followings:**
   Inoculum:

   Pure culture:

   Aseptic technique:

   Sterile medium:

   Colony:

4. **Define and explain the use of the following media:**
   Broth:

   Agar:

   Slant:

   Stab agar:

   Petri plate:

   Semi-solid media:

5. **Why an agar is a good ingredient in the conversion of liquid media to solid media.**

6. **How can autoclave temperature go above the boiling temperature of water?**

7. **Discuss the important points in the preparation of media and solution preparation in the laboratory.**

8. **Why wet heat is more efficient than dry heat as a sterilizing agent?**

9. **Why is the Petri plate inverted during incubation or storage?**

10. **What is the primary use of slants and stabs?**

PRACTICE 2

Microorganisms in environment

2.1 Theoretical information

Microorganisms, bioaerosols, air conduction, and biological materials are distributed to the environment by accumulation in the gas environment. Bioaerosols are sneezing, coughing, wave motion, splashes, wind, cooling towers, ventilation systems, air circulations, etc. They are formed by a wide variety of natural and man-made processes. Inhalation, swallowing, and skin contact are ways of human exposure to microorganisms, viruses, and similar agents in the air, but breathing is the most important route that has a negative impact on human health. Pollen, molds, bacteria, spores, viruses, and others in the air are transmitted to people by aerosols. Disruption of building materials and spoilage of food may be associated with microbial contamination of indoor environments such as residences, offices, schools, health facilities, closed agricultural structures (stables and crop storage areas), and industrial facilities.

In the detection of microorganisms from environment, various sampling techniques from air are used. Filtration technique (such as membrane filter) can be used largely for the collection of microorganisms from air. The filtrate on the filter paper concentrates the microorganisms onto a smaller area. Filter is microscopically examined by culturing on solid media. Microbial count from air can also be detected using direct exposure of agar media to air or exposure of plate to a volume of air flow. In this practice, microorganisms will be detected and counted from air and people using various techniques.

2.2 Microbial detection techniques from environment

Materials

Media and solutions. Nutrient broth, nutrient agar (NA) slant, NA plate, potato dextrose agar (PDA) plate and Sterile 300 mL of 0.1% peptone water in 500 mL flask.

Equipment. Loop, antifoam, sterile graduated cylinder (100 mL), sterile pipettes (1 and 10 mL) in box, forceps, pipette discard pod containing disinfectant solution (0.25% hypochlorite), Bunsen burner, vortex, and impingement device.

https://doi.org/10.1016/B978-0-323-91017-0.00018-4

2.2.1 Exposure of petri plate to air

1. Invert the plates, label the bottom with the time of exposure and your name or initials (Fig. 2.1).
2. Remove the cover of a Petri plate (NA and PDA plates).
3. Expose the uncovered plate to air for specified times (0, 10, 20, and 30 min).
4. Replace the cover of plates after exposure time.
5. Incubate inverted plates in an incubator at 35°C for 24 h.
6. After incubation, count colonies from plates and record results for the followings: appearance, size, form, elevation, optical properties, and color of colonies.

2.2.2 Contamination of microorganisms from student

1. Label the bottom of the NA and PDA plates with name of medium, your initials.
2. Remove the cover of NA and PDA plates and expose one plate (from NA and PDA) to each of the following directions:
   - **(a)** touching the surface of the agar medium in the plate with your fingertips,
   - **(b)** contact your hair over second agar medium in the plate,
   - **(c)** coughing over third agar medium in the plate and.
   - **(d)** place a coin (money) over fourth agar medium in the plate and remove coin using disinfected forceps (dipping into alcohol and flaming).
3. Replace the cover of plates.
4. Incubate inverted plates in an incubator at 35°C for 24 h.
5. After incubation, count colonies and record results for the followings per plate: number, size, and color of the colonies.

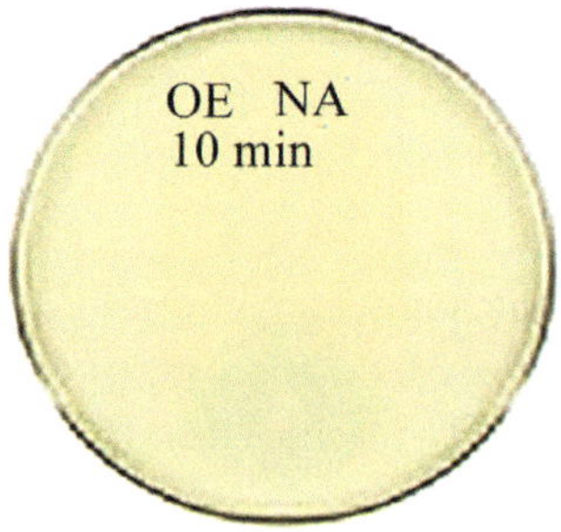

Fig. 2.1 Labeled Petri plate.

2.2.3 Air sampling by impingement

1. Sterilize all parts of impingement devices in the autoclave by placing distilled water into glass and closing all openings with aluminum foil.
2. Add 20 mL of sterile 0.1% peptone water into impringer, add 0.1 mL of sterile anti-foam and label the water level on the impringer.
3. Set up air sampling impingement as shown in Fig. 2.2.
4. Turn on the pump (vacuum source) for 5, 10, 20, and 30 min. Open vacuum for a next time on same liquid to complete time for next period. Remember, the water volume in the impringer decreases during impingement because of evaporation and sample removal. Transfer after each time, add sterile water up to labeled part of impringer.
5. With the 1-mL pipette, remove 1 mL of fluid from the impringer after each exposure time. Place 1-mL sample onto each of NA and PDA plates.
6. Spread plate the sample under aseptic conditions (as indicated in Practice 4). Replace used pipette into pipette discard pod.
7. Incubate the inverted NA plates at 35 °C for 24 h in the incubator and SDA plates at 25 °C for 2–5 days in the incubator.
8. After incubation, count colonies, and examine and record results for the followings per plate: number, size, and color of colonies. Record number of colonies for bacteria, yeasts, and molds.
9. Calculate number of bacteria from NA plate and fungi from PDA plate per liter of air. The air passing amount water (liquid) may be 12.5 L per minute. The microorganism is usually reported as numbers per liter of air, which is calculated as follows:
   - **(a)** Volume of air (L) = Air sampling time (min) X air flow rate/min.
   - **(b)** Number of microorganisms (cfu/mL) in liquid = count (cfu/mL) X mL liquid remaining in impringer.
   - **(c)** Number of microorganisms (cfu/L of air) = Number of microorganisms (cfu/mL)/Volume of air (L).

Fig. 2.2 Collection of biological aerosols from air in liquid using impingement device.

LABORATORY REPORT

Date:

Name:

Lab Title: Microorganisms from Environment

(1) Record followings from air exposure of Petri plates:

(a) For NA plates

Number of colonies per plate:

Number of colonies according to color per plate:

Number of colonies according to size per plate:

(b) For PDA plates

Number of colonies per plate:

Number of colonies according to color per plate:

Number of colonies according to size per plate:

(2) Record colony observations (color, size, number) from:

Hand:

Hair:

Coughing:

Money:

(3) Average number of colonies from air sampling using imprringement:

Bacteria:

Yeasts:

Molds:

(4) Calculate bacteria and fungi (molds and yeasts) numbers per liter of air.

(5) Which factors can increase number of microorganisms in air?

(6) Which types of microorganisms are commonly associated with air and your skin?

(7) Discuss the importance of microorganisms in air.

PRACTICE 3

Aseptic techniques

3.1 Theoretical information

Aseptic techniques (asepsis) are the prevention of microbial contamination from working air, equipment, surface area, and humans. Aseptic transfer of a culture is successfully performed without any microbial contaminants. Some of the basic aseptic technic providing elements are hand washing, antiseptic, disinfectant, Bunsen burner flame, etc.

Hand washing. Before start working in the laboratory, wash hands with a liquid detergent, soak them in antiseptic solution (such as 0.5% zephran) for 10 min, wash your hands again with water and dry them with a paper towel. After completing the practice, wash your hands and apply antiseptic again waiting 10 min and wash your hands again before leaving the laboratory.

Antiseptics. Antiseptics are antimicrobial agents that are used externally on living tissues (the skin and out of the cavities to the mucous membrane) for the destruction or further reducing microorganisms and viruses. Antisepsis is the application of chemicals locally to the body surface to kill microorganisms or prevent them from multiplying. Antiseptics are used in the treatment of skin infections; wounds, cuts, and abrasions; in the protection of the skin at the end of the surgery; and in cleaning of the hand before surgery. Some of the antiseptics are alcohols (70%), quaternary ammonium compounds (zephran), chlorhexidine hexachlorophene, hypochlorite, iodine, peroxides, permanganates, phenol derivatives, quinolone derivatives, etc.

Disinfectants. These are chemicals (0.1% hypochlorite, 70% alcohol, or 0.5% zephran) that can be used to kill any microorganisms that may be present on the surface of the working area at the beginning of the practice. Cleaning with disinfectants can be done by adding disinfectants onto the surface area that is effective on vegetative cells of microorganisms. Disinfectants are not effective on spores. Disinfections also remove existing dust and minimizes the possibility of microbial contamination. Effects of disinfectants according to the structures and components on microbial cells are given in Table 3.1.

Laboratory Practices in Microbiology
https://doi.org/10.1016/B978-0-323-91017-0.00011-1

Table 3.1 Effects of disinfectants according to the structures and components on microbial cells.

Cell membrane	Cell proteins	Protein and nucleic acid	Enzymes	Bacterial spores
Ionic surfactant (soap, sodium dodecyl sulfonate) Non-ionic surfactant: Phenol and its derivatives (cresol, lysol, hexachlorophene) and organic solvents (chloroform, alcohol)	Alcohol, acetone, organic solvents	Heavy metals (mercury, zinc, cupper) Oxidizing agent (peroxides) Alkaline agents (formalin and ethylene oxide)	Zephran, oxidizing compounds, formaldehyde, ethylene oxide	Phenol, glutaraldehyde, formaldehyde, hypochlorite, iodine, hydrogen peroxide, ethylene oxide

Loop and needle. Microorganisms can be transferred from one media to another using a loop or needle (Fig. 3.1b). The loop and needle should be heated through Bunsen burner flame (Fig. 3.1b) until they become red before use and after inoculating the microorganism. Before contact of loop or needle with the culture to be taken, it should be allowed to cool enough near flame to prevent the killing of microbial cells to be transferred. If excess culture is to be transferred, then loop should be used in the transfer. Needles should be preferred in pure culture isolations and when transferring culture into solid media. In pure culture isolation, culture should be taken by touching on the center of the colony with a needle or loop.

Bunsen burner use. Aseptic conditions can be achieved through the Bunsen burner flame during microbial transfers (Fig. 3.1b). Bunsen burner should be closed when not in use. Temperature in a small laboratory room increases with the excessive use of Bunsen burners. More importantly, the Bunsen burner flame becomes dangerous for hair, clothing, and skin.

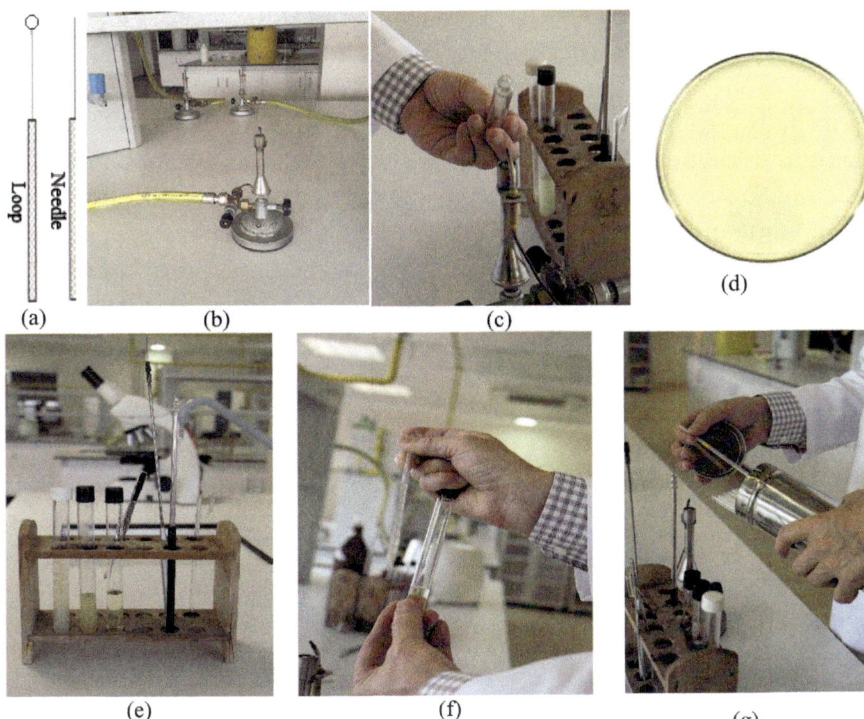

Fig. 3.1 Laboratory equipment: loop and needle (a), Bunsen burner (b), flaming the mouth of tube (c), Petri plate (d), receptacle (e), removing cap of tube, (f) and removing pipette from pipette box (g).

Test tube. The cap of the tube used in a practice is removed by rotating the tube (Fig. 3.1f) first and the mouth of the tube is flamed (Fig. 3.1c). After the use of the tube, the tube mouth is re-flamed and the cap is closed by rotating the tube. The tube is placed to receptable (Fig. 3.1e). The cap of the tube should never be left on the desktop during inoculation.

Inoculation to medium in tube. If a liquid medium in the tube is to be inoculated with the microbial culture, then the mouth of tube must be flamed before the needle or loop is inserted into the tube. The needle or loop must be moved forth and back in the fluid to allow the microorganisms pass to liquid and mixing. After inoculation, the loop or needle is removed from the tube, flamed as before, and returned to a receptacle. The inoculated tube should be passed through the flame and then lid should be closed. The loop or needle should be burned after inoculation and should never be left on the table.

Inoculation onto medium in Petri plate. During inoculation into the medium in the Petri plate (Fig. 3.1d), a loop or a sterile pipette is used in this inoculation. When the microorganisms are taken from the solid culture in a Petri plate, the cover of plate should only be partially opened under aseptic conditions (near the Bunsen burner flame) and inoculation is made by opening plates quarter to prevent contamination of microorganisms during spreading the inoculum onto the agar medium. When the microorganisms are taken from liquid culture, take a sterile 1 mL pipette by flaming the mouth of pipette box and pipette tips, and remove pipette near flame (Fig. 3.1g). Remove 1 mL of culture and add on agar medium in the Petri plate. In order to prevent contamination of microorganisms with the medium in the Petri plate, its cover should always be kept closed.

Disposal materials. The rules indicated in the followings can be applied to discarded culture, used glassware, and broken glass disposal:

1. Place Petri dishes in a bag to be sterilized.
2. Place unused cultures in the tubes into a wire basket to be autoclaved.
3. Place used pipettes in a pipette bag for autoclaving. The bag of pipette should contain a disinfectant (such as 0.25% hypochlorite). Place calico stopper into bag to prevent breaking tip of pipette.
4. Sweep the broken glasses into a dustpan and put them in a container reserved for them. Do not try to pick up the glass fragments with your fingers.
5. Never place contaminated material in a wastebasket in the laboratory.
6. When culture drops onto desk top, pour an amount of disinfectant onto drops, allow for 15 min and use a sponge to scrubbing it clean.

3.2 Types of disinfectants and antiseptics

3.2.1 Compounds used as high-level disinfection

Glutaraldehyde. Glutaraldehyde (a saturated dialdehyde) is an effective disinfectant and chemical sterilizer. Its acidic solution does not have sporicidal properties. When used only as an alkaline solution (pH 7.5–8.5), it has a sporicidal property. Its 2% solution can kill vegetative bacteria in 2 min, mycobacteria in 20 min, and spore forming bacteria in 3 h. It is used as a sterilant for 3 h. It is not corrosive for metal and does not damage plastic and rubber material.

Orthofitalaldehyde. It is used as disinfectant with the 0.55% solution at pH 7.5. There is no irritant effect on eye and nose passage. It has a wide range of effects on microorganisms and spores. It is less effective than glutaraldehyde and faster to mycobacteria.

Peracetic acid. Peracetic acid or peroxyacetic acid is a sterilizer that influences all microorganisms and spores. Its 0.2% solution is effective on microorganisms within 5–10 min and used as sterilant. It is a durable and expensive compound. It also causes corrosion on some metals. Skin burns and eye damage may occur in contact with intensive solution.

Hydrogen peroxide (H_2O_2). It is effective against microorganisms, bacteria, fungi, viruses, and spores. Its 7.5% solution can be used as sterilant with 20 min exposure. Its 3% solution is used as disinfectant for 10 min. It is a non-toxic compound.

Chlordioxide (ClO_2). It is a highly oxidizing substance. It is corrosive and irritant. It is affected by organic substances and light. Five min exposure is enough for high degree of disinfection.

3.2.2 Compounds used as medium and low-level disinfection

Alcohols. They are effective against Gram-positive and Gram-negative bacteria, mycobacteria, fungi, and various viruses such (as rotavirus, rhinovirus, coronavirus, herpes simplex virus, influenza virus, hepatitis B virus, HIV, adenovirus, respiratory syncytial virus). They are not recommended for surface cleaning as they evaporate. Ethyl-isopropyl alcohol solution (60%–90%) effective on microorganisms within 15 min (such as 70% alcohol + 5% isopropyl alcohol + 0%–10%alkyl-dimethyl-ammonium chloride).

Quaternary ammonium compounds. Benzalkonium chloride, cetyl-pyridinium chloride, and alkyl-dimethyl-ammonium chloride are disinfectants in this group. Generally, it is an effective disinfectant on the surface that shows cationic detergent feature. It can also be used for cleaning the surfaces of non-critical instruments, such as blood pressure cuff and stethoscope. Its 0.5% solution effective on microorganisms within 10 min. Alkyl dimethyl benzyl ammonium chloride is the main compound used in hospitals. It has activity as fungicide, bactericide, and virucide.

Phenolics. They are phenol derivatives used in high concentrations, such as orthophenylphenol and ortho-benzyl-para-chlorophenol. It is used as fungicides, tuberculosides, and viruses. It is recommended for use in cleaning non-critical medical devices.

Iodophores. Iodine and polyvinyl pyrrolidone compounds are the most used iodophor and has a bactericidal, tuberculocidal, viricidal, and fungicidal effect. Their 30–150 ppm solution effective against these microorganisms within 15 min. They do not have sporicidal effects. They are often used as an antiseptic, as well as for cleaning blood culture bottles and medical equipment, thermometers, and endoscopes after use.

Hypochlorite. It has a wide spectrum of action as bactericide, fungicide, virucide, and parasite, and is used in household cleaning, dairy industry, and swimming pools. Bleach (NaOCl) is sodium hypochlorite at a concentration of 5.25%. Depending on the amount of organic matter on the surfaces to be disinfected, it is diluted up to 1/10–1/100. 1/10 dilution of 5.25% stock solution of sodium hypochlorite is equivalent to approximately 5000 ppm free chlorine. This solution is recommended by the CDC (Centers for Disease Control) for disinfection of blood spilled on the ground and splashing around. For very clean surfaces, 1000 ppm active chlorine is sufficient. Its 5% solution is used as disinfectants for 1 min on places and horizontal surfaces. Its 50–5000 ppm solution effective on microorganisms within 15 min. The disadvantages are that they are corrosive (metals), inactivated in the presence of organic matter and are not durable.

3.2.3 Culture media used in aseptic transfers

Broth culture is a liquid medium in tube containing microbial cells. Tubes containing the broth are inoculated with microbial culture and incubated at optimum growth temperature. After incubation of liquid medium in tube, growth of microorganisms is observed as one or a combination of three forms (Fig. 3.2a):

Pellicle: A mass of microorganisms is floating on top of the broth.

Turbidity: The microorganisms appear as a general cloudiness throughout the broth.

Sediment: A mass of microorganisms appears as a precipitate at the bottom of the tube.

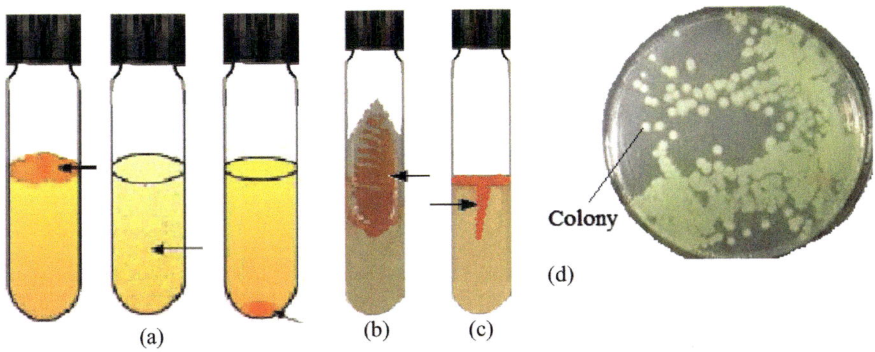

Fig. 3.2 Types of bacterial growth (a) in broth cultures (from *left* to *right*: pellicle, turbidity and sediment), (b) on slant culture, (c) on stab culture, and (d) on plate culture. *Arrows* indicate microbial growth.

Slant culture is the growth of microorganisms on the surface area of agar slant. The agar slant is inoculated with microbial culture and incubated at optimum growth temperature of microorganism. The microorganism produces colony on the surface area of agar slant (Fig. 3.1b).

Stab culture (deep) is the agar medium containing microbial growth through inoculation line of agar and surface (Fig. 3.2c) of agar medium in tube. The medium is inoculated with microbial culture by inserting wire needle into agar medium and incubated at optimum growth temperature of microorganism.

Agar plate culture is Petri plate containing microbial colonies. Petri plate is inoculated by microbial culture (such as by streaking) and incubated. The microorganisms produce colonies on or through the medium (Fig. 3.2d).

In order to perform aseptic transfers of microbial cultures, you need to gain some application skills. In this practice, five sample transfer techniques will be made: (a) liquid culture to broth, (b) liquid culture to agar slant, (c) plate culture to agar slant, (d) plate culture to broth, (e) slant culture to broth, and (f) broth culture to stab agar.

3.3 Aseptic transfer techniques in microbiological researches

Materials

Microbial cultures. 24 h nutrient broth, slant, and agar plate cultures of *Escherichia coli*.

Media. Nutrient broth in tube, nutrient agar slant, nutrient agar deep, and nutrient agar plate.

Equipments. Loop, needle, Bunsen burner, disinfectant, and water-proof marking pen.

3.3.1 Transfer from liquid culture into broth

1. With a water-proof marking pen, label a tube of sterile nutrient broth with your initials, name of microorganism and type of transfer (liquid culture to broth).
2. Take the loop or needle from receptacle (Fig. 3.3a) by hand (Fig. 3.3b). Sterilize loop by holding it at 90 degree angle through upright over the flame of the Bunsen burner until all the wire becomes redness (Fig. 3.3c). The entire wire must be heated and cool the loop for at least four seconds near flame.
3. Using free hand, take the liquid culture tube from receptacle (Fig. 3.3a) and gently shake the tube to disperse the culture (Fig. 3.3d).
4. Grasp the tube cap with your little finger of hand while holding the inoculating loop and remove cap of the tube by rotating the tube counter clockwise (Fig. 3.3e).
5. Flame the tube mouth (Fig. 3.3f).
6. Insert the loop into the culture (Fig. 3.3g) and remove a loopful of culture.
7. Re-flame the tube mouth again (Fig. 3.3h), replace the cap of tube by turning the tube clockwise (Fig. 3.3i) and place the tube back on the receptacle (Fig. 3.3a).
8. Take a nutrient broth the tube with your free hand, carefully remove the cap of tube with little finger of your hand by rotating the tube counter clockwise while holding the inoculating loop (Fig. 3.3j).
9. Flame the tube mouth (Fig. 3.3k).
10. Insert the loop into the sterile broth in the tube and disperse the microorganisms in the broth (Fig. 3.3l), move the loop back and forth in the tube in a short high.
11. Remove the loop from the tube and re-flame the tube mouth (Fig. 3.3m).
12. Replace the cap on the tube by rotating the tube clockwise (Fig. 3.3n).
13. Place the inoculated broth on the receptacle (Fig. 3.3a).
14. Sterilize the loop by flaming (Fig. 3.3o). Return the loop to receptacle (Fig. 3.3a).
15. Incubate the nutrient broth tube you just inoculated at 35°C in an incubator for 24 h.
16. After incubation, examine tubes for growth (turbidity) (Fig. 3.2a) and record results.

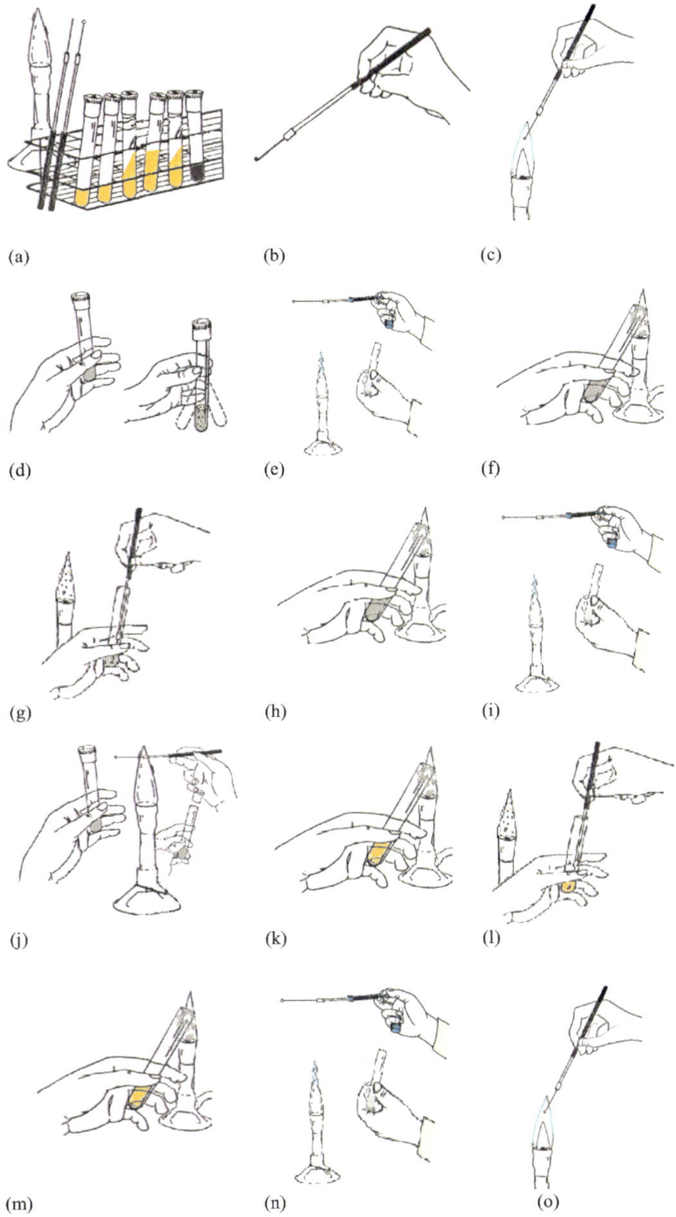

Fig. 3.3 Transferring microorganisms from a liquid culture into a nutrient broth. (a) Receptacle, loop, Bunsen burner flame, solid and liquid media in tubes, and liquid culture in tube. (b) Hold loop in one hand. The loop should be held like a pen. (c) Flame the loop in the flame while holding the loop 90 degree angle so that all of the nichrome wire gets red hot. (d) Hold tube in another hand and the microbial culture is dispersed by gently shaking the tube. (e) Remove cap from culture tube. Do this by wrapping little finger of your hand. (f) Flame mouth of tube by passing back and forth through flame twice in a nearly horizontal position. (g) Insert loop into culture and remove a loopful of culture. (h) Re-flame mouth of tube again. (i) Recap cap on tube. (j) Take the tube and remove cap of tube. (k) Flame mouth of tube. (l) Inoculate a loopful of culture into broth. *(Modified with permission from McGraw-Hill; Benson, H.J. Microbiological applications: laboratory manual in general microbiology, 2001, Fig. 21.3, p. 61).*

3.3.2 Transfer from liquid culture on to agar slant

1. Follow the steps 2–7 as explained in procedure "3.2.2".
2. Take a nutrient agar slant tube with your free hand, carefully remove the cap of tube with little finger of your hand by rotating the tube counter clockwise while holding the inoculating loop (Fig. 3.3j).
3. Flame the tube mouth (Fig. 3.3k).
4. Inoculate culture by placing loops onto slant surface starting by short streak from bottom of the tube (Fig. 3.4). Move loop upward out of the tube (not until upper end of agar medium). Do not dig into the agar.
5. Re-flame the mouth of the agar slant tube (Fig. 3.3m) and replace the cap of the tube by rotating the tube clockwise (Fig. 3.3n).
6. Place the inoculated agar slant tube on the receptacle (Fig. 3.3a).
7. Flame the loop (Fig. 3.3o) and place it on the receptacle (Fig. 3.3a).
8. Label the tube with the name of microorganism, the date, your name, or initials and type of transfer.
9. Incubate the nutrient agar slant tubes at 35°C for 24 h in the incubator.
10. After incubation, examine slant in the tube for colony formation as growth (Fig. 3.2b) and record results.

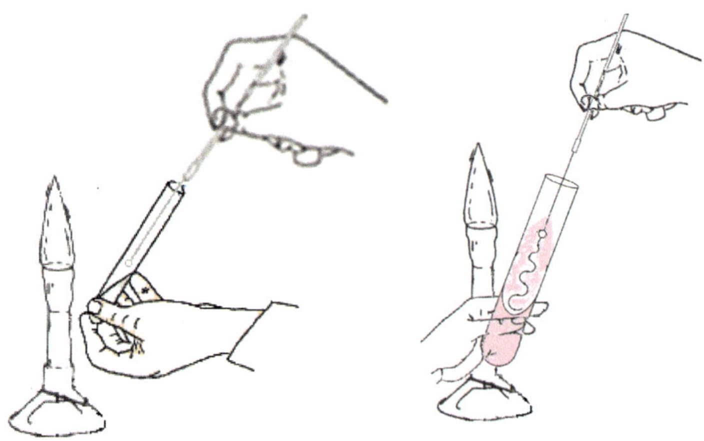

Fig. 3.4 Short streak on the surface area of slant starting from bottom to upward.

3.3.3 Transfer from plate culture (by colony selection) to agar slant

1. Select a colony on the nutrient agar plate and draw a circle around it with a water-proof marking pen on underside of the Petri plate (Fig. 3.5a).
2. Take the loop with one hand (Fig. 3.3b) and flame the wire portion of loop to redness (Fig. 3.3c) and cool the loop for at least 4 s near flame.
3. Lift Petri plate lid top (Fig. 3.5b), cool the loop by touching on the clear agar surface and do not completely remove the plate lid to exposing the agar surface to the air.
4. Remove inoculum by touching the loop onto the center of the selected colony (Fig. 3.5b). Do not dig into the agar with your loop during pick up microorganisms.
5. Close the lid of plate.
6. Take a nutrient agar slant from receptacle (Fig. 3.3a), remove cap of tube by rotating the tube counter clockwise while holding the inoculating loop (Fig. 3.3j).
7. Flame the tube mouth (Fig. 3.3k).
8. Place the loop onto the slant surface starting from the bottom of the tube (Fig. 3.4). Move loop from side as you pull it upward out of the tube (not until upper end of agar medium). Do not dig into the agar.
9. *Re*-flame the mouth of the agar slant tube (Fig. 3.3m), replace the cap of the tube by rotating the tube clockwise while holding the inoculating loop (Fig. 3.3n) and place it on the receptacle (Fig. 3.3a).
10. Flame the loop (Fig. 3.3o) and place it on the receptacle (Fig. 3.3a).

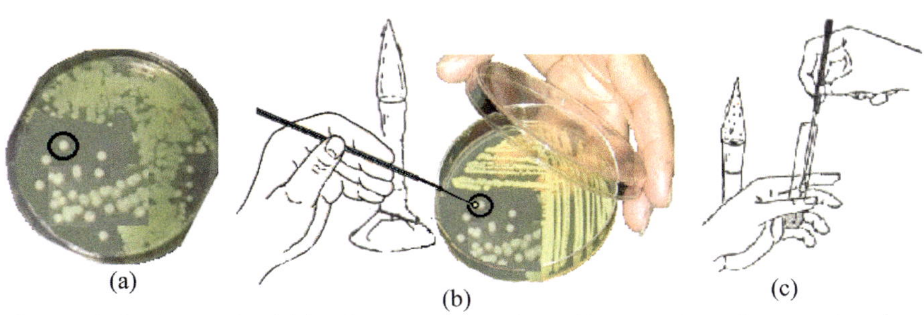

Fig. 3.5 Selecting a microbial colony on plate culture (a), removing microorganism from colony (b), and inoculating into broth (c).

11. Label tube with name of microorganism used, the date, your name or initials, and type of transfer.
12. Incubate inoculated nutrient agar slant tubes at 35°C in an incubator for 24 h.
13. After incubation, examine the slant in the tube for growth (Fig. 3.2b) and record your results.

3.3.4 Transfer from plate culture (by colony selection) to broth

1. Follow the steps 1–5 as explained in procedure 3.3.4″ (but use needle).
2. Take a nutrient broth from receptacle (Fig. 3.3a) and remove the cap of the tube by grasping them with the little finger of the hand by rotating the tube counter clockwise while holding the needle (Fig. 3.3j)
3. Flame mouth of the tube (Fig. 3.3k).
4. Insert the needle into the sterile broth and inoculate culture (Fig. 3.3l). To disperse the microorganisms in the liquid medium, move needle back and forth in the tube in a short height.
5. Flame the tube mouth (Fig. 3.3m), replace the cap of the tube by rotating the tube counter clockwise while holding the needle (Fig. 3.3n) and place it on the receptacle (Fig. 3.1c).
6. Flame the needle (Fig. 3.3o) and place it on the receptacle (Fig. 3.3a).
7. Label the tube with the name of microorganism, the date, your name or initials, and type of transfer.
8. Incubate inoculated nutrient broth tubes at 35°C in an incubator for 24 h.
9. After incubation, examine broth culture in the tube for growth (turbidity) (Fig. 3.2a) and record your results.

3.3.5 Transfer from slant culture to broth

1. Take the needle with hand (Fig. 3.3b) and flame the wire portion of needle to redness (Fig. 3.3c). Cool the needle at least 4 s near flame.
2. Takes a nutrient agar slant culture (Fig. 3.2b), remove the cap of the tube by grasping them with the little finger of the hand by rotating the tube counter clockwise while holding the needle (Fig. 3.3e).
3. Flame mouth of the tube (Fig. 3.3f).
4. Remove an inoculum by touching needle onto the center of the colony (Fig. 3.6a). Do not dig into the agar.
5. Re-flame the mouth of tube (Fig. 3.3h), replace the cap of the tube by rotating the tube clockwise while holding the needle (Fig. 3.3i) and place the tube on the receptacle (Fig. 3.3a).

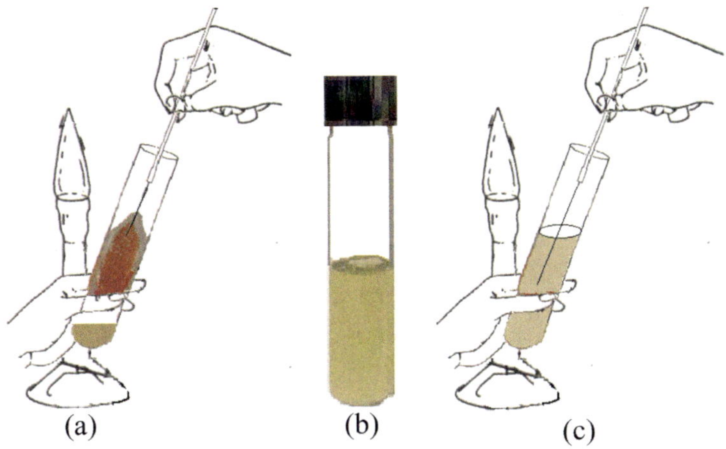

Fig. 3.6 Removing culture from agar slant (a). Inoculation of culture into agar deep (b and c).

6. Take the nutrient broth tube (Fig. 3.2j), remove the cap of the tube by grasping them with the little finger of the hand by rotating the tube counter clockwise while holding the needle.
7. Flame mouth of the tube (Fig. 3.3k).
8. Insert the needle into the broth and inoculate it (Fig. 3.4l). To disperse the microorganisms in the liquid medium, move the needle back and forth in the tube in a short high.
9. Re-flame the tube mouth (Fig. 3.3m) and replace the cap of the tube by rotating the tube clockwise while holding the needle (Fig. 3.3n).
10. Place the inoculated tube on the receptacle (Fig. 3.3a).
11. Flame the loop (Fig. 3.3o) and place it on the receptacle (Fig. 3.3a).
12. Label the tube with the name of microorganism, the date, your name or initials, and type of transfer.
13. Incubate inoculated nutrient broth tube at 35°C in an incubator for 24 h.
14. After incubation, examine broth culture in the tube for growth (turbidity) (Fig. 3.2a) and record your results.

3.3.6 Transfer from broth culture to agar stab

1. Take the needle with hand (Fig. 3.3b), flame the wire portion of needle to redness (Fig. 3.3c), and cool the loop for at least four seconds near flame.
2. Take a broth culture with another hand (Fig. 3.3d) and remove the cap of the culture tube by grasping them with the little finger of the hand by rotating the tube counter clockwise while holding the needle (Fig. 3.3e)

3. Flame the mouth of tube (Fig. 3.3f).
4. Obtain an inoculum by inserting the needle into the broth culture (Fig. 3.3g).
5. Re-flame the tube mouth (Fig. 3.3h), replace the cap of the tube by grasping by rotating the tube counter clockwise while holding the needle (Fig. 3.3i) and place the tube back on receptacle (Fig. 3.3a).
6. Take the nutrient agar stab tube (Fig. 3.6b) and remove the cap of the tube by grasping them with the little finger of the hand by rotating the tube counter clockwise while holding the needle (Fig. 3.3j).
7. Flame the mouth of tube (Fig. 3.3f and k).
8. Inoculate culture to the deep agar by inserting needle into agar in the tube (not up to bottom of the tube) (Fig. 3.6c).
9. Re-flame the mouth of tube (Fig. 3.2m), replace the cap of the tube by rotating the tube clockwise while holding the needle (Fig. 3.3n).
10. Place the tube on receptacle (Fig. 3.3a).
11. Flame the loop (Fig. 3.3o) and place it on the receptacle (Fig. 3.3a).
12. Label the tube with the name of microorganism, the date, your name or initials, and name of transfer.
13. Incubate inoculated nutrient agar stab tubes at 35°C in an incubator for 24 h.
14. After incubation, examine tubes for the growth (Fig. 3.2c) and record your results.

LABORATORY REPORT

Date:
Name:

Lab Title: Aseptic Techniques

1. **Observe the growth in the following cultures with respect to growth, pigmentation, and purity**

   Liquid culture inoculated into broth:

   Liquid culture inoculated onto agar slant:

   Plate culture inoculated onto agar slant:

   Plate culture inoculated into broth:

   Slant culture inoculated into broth:

   Broth culture inoculated onto stab agar:

2. Define pure culture:

3. Why must the loop or needle be flamed before and after removing a culture?

4. Why do you perform these transfer techniques?

5. Why is it necessary to sterilize the loop between streaks when streaking for single colonies?

6. Why is the aseptic transfer so important?

7. Why is the mouth of flask or test tube flamed while making transfer of cultures?

PRACTICE 4

Pure culture techniques

4.1 Theoretical information

For microorganisms to grow and identify, inoculation into laboratory media involves basic and important pure culture techniques. Extreme care must be performed in pure culture technique in order to prevent contamination. The inoculation into a medium is carried out by using aseptic techniques. Aseptic techniques prevent microbial contamination with media and microbial culture from environment, such as air, surface area, hands, respiration, and other nonsterile sources. Several aseptic precautions are used in the pure culture researches. The working instruments (forceps, loops, needle, etc.), and the rims of bottles and other vessels are sterilized before and after works by flaming. The contamination risk of microorganisms with culture and media is further reduced by treating the benches surface area with disinfectants or UV (ultraviolet light). In some cases, filtering the ambient atmosphere prevents enter of the microorganisms, their spores, and other matter to working room.

The microbial culture transferred to the microbiological medium is called the inoculum. During incubation, individual cells multiply in medium; they produce a higher number of cells that together form a colony (visible to the naked eye) on the agar medium or large number of cells form turbidity in the liquid medium. Single colony on the agar medium indicates a pure culture formation from a single microbial cell (a colony coming from a single cell forming daughter cells). A pure culture (here colony) consists of only one type of microorganism. The characteristics of microorganisms cannot be studied in the mixed culture. Several different techniques of getting a pure culture from a mixed culture are available. The most frequently used pure culture techniques are streak plate, spread plate, and pour plate.

1. Pour-plate technique. In this technique, a known volume of culture, usually 1 mL, is added into a sterile Petri dish by sterile pipette; melted agar medium is then poured into Petri dish and mixed. The plates can solidify and then are inverted to prevent moisture drop onto agar surface. Growth of microorganisms after incubation occurs both on the surface area and the depths of the agar medium. The initial step of pour-plate method is

Laboratory Practices in Microbiology
https://doi.org/10.1016/B978-0-323-91017-0.00002-0

the preparation of serial dilutions of sample in sterile diluent (such as 0.85% NaCl and 0.1% peptone water) to provide colony formation on plates from 30 to 300.

2. Spread-plate technique. The spread-plate is another technique for obtaining the pure culture. A known volume of culture, usually 1 mL, is added onto sterile Petri plate and spread over the surface area of plate. The plates are inverted to prevent moisture drop onto agar surface. Growth of microorganisms after incubation occurs on the surface area of the agar medium. The initial step of pour-plate technique is the preparation of serial dilutions from sample in sterile diluent (such as 0.85% NaCl and 0.1% peptone water) to provide colony formation on plates from 30 to 300.

3. Streak-plate technique. The streak-plate involves the streaking of microbial culture onto the surface of the agar medium in the plate by a loop. A series of streaks spreads the inoculum on the agar medium. The agar surface should be free of condensation, since the presence of moisture on the surface of the medium would interfere with the formation of separate colonies. The streak plate is used to obtain single colonies as pure culture. Growth occurs after incubation on the surface of the agar medium.

Besides obtaining single colonies, pour and spread plate techniques are also used to count the number of viable microorganisms from foods and cultures. There are several disadvantages in the use of plate techniques: (1) colonies of several microbial species/strains may present with a similar appearance on/in the agar, (2) some bacterial species may not grow on/in the medium (with or without oxygen), (3) microorganisms can be inactivated by hot medium (over 45°C) in pour plate, (4) in order to carry out further studies in pour plate method, it may be difficult to remove colonies from inside of the medium, and (5) they require more diluents, media, tubes, pipettes, and plates.

4.2 Plate culture techniques

The surface of your table top should be disinfected with disinfectant (Rocal, Zephran, Betadine or hypochlorite, etc.) in order to ready for experimental researches. Use a sponge to scrub it clean.

Materials

Microbial cultures: 24 h nutrient broth mixed cultures of *Escherichia coli* and *Staphylococcus aureus.*

Media and solutions: Nutrient agar plates, nutrient agar stab (20 mL in tubes) and sterile 0.1%) peptone water (9 mL).

Equipments: Water-bath (at 60°C) to melt agar stab, sterile Petri dishes, sterile pipettes (1 mL) in box and pipette discard pod containing disinfectant (0.25% hypochlorite) and water proof marking pen.

4.2.1 Pour-plate technique

1. Before beginning the procedure, label the bottom of the sterile Petri dishes, name of microorganism, dilution factor, pour-plate name, and your initials using water-proof marking pen.
2. Place the nutrient agar stabs in the water bath (at 60°C) and allows the contents to melt evenly, retain the melted agar medium at 48°C (or cool sterilized medium in flask to 48°C).
3. Make successive 10 fold dilutions from bacterial culture. Transfer 1 mL sample into 9 mL sterile diluent. The quantitative dilution and inoculation into sterile Petri dishes are given in Fig. 4.1.
4. For this inoculation, remove 1 mL pipette from box (Fig. 4.2a). Open pipette box slightly and move the pipettes up to mouth of box by tilting. Remove lid and flame the mouth of box and pipettes. Take a pipette, flame the mouth of box again, and place the cap of box.
5. Mix the sample in diluent using vortex mixer (Fig. 4.2b) or rotating through hand (Fig. 4.2c).
6. At the same time of dilution procedure, remove 1 mL diluted or non-diluted sample using the sterile pipette (Fig. 4.3), transfer 1 mL into slightly opened sterile Petri dish in duplicate under aseptic conditions (Fig. 4.3c).

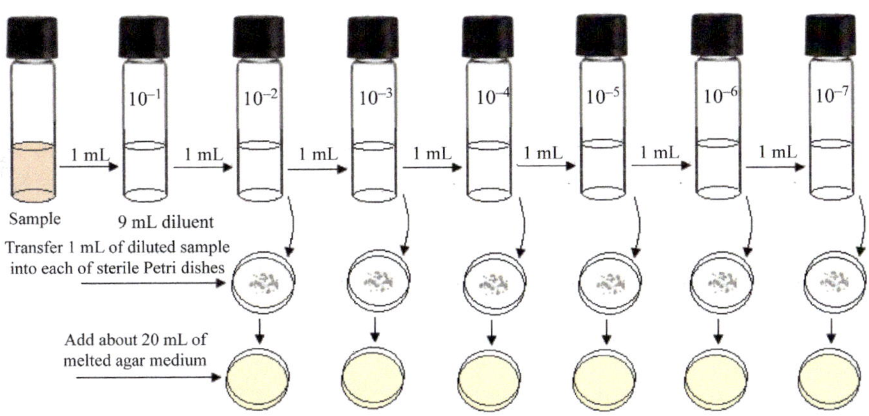

Fig. 4.1 Sample dilution, transferring diluted sample into sterile Petri dishes and pouring melted agar medium into them.

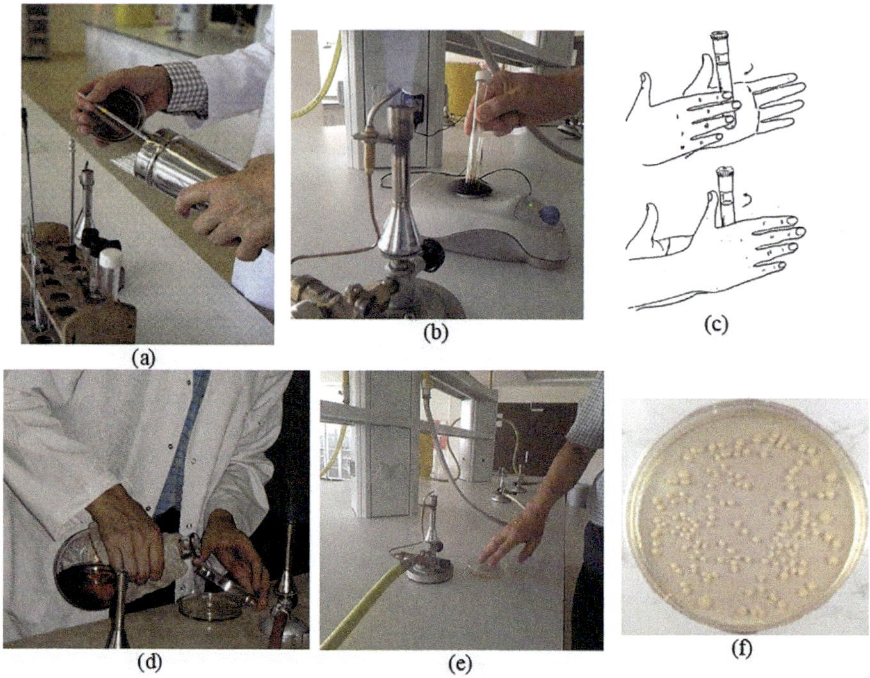

Fig. 4.2 (a) Remove a sterile pipette from box, (b) vertexing the tube to mix the sample in the diluent [or (c) rotate the tube vigorously between hands approximately 10 times to mix the sample in the diluent. Take care not to let any of the sterile distilled water splashes onto closures], (d) pour about 20 mL cooled sterile medium into sterile Petri dishes under aseptic condition, (e) mix the inoculum and medium by rotating the plates in the route drawing 8 while it is lying on the table top, and (f) colony formation on plates from 10^6 dilution.

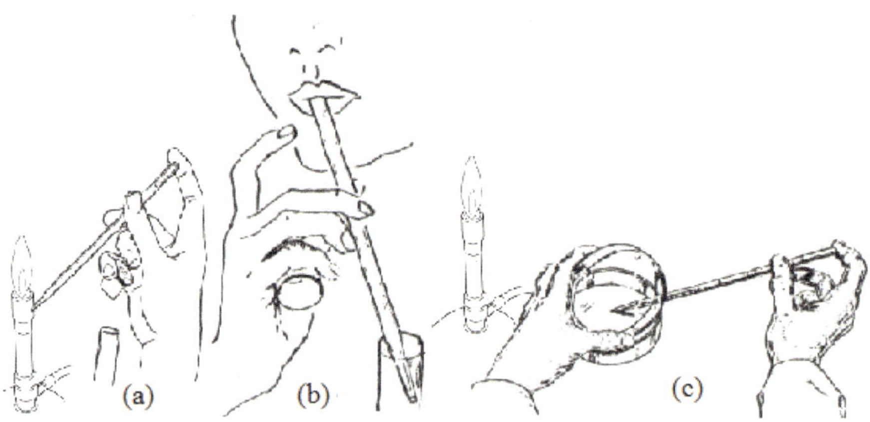

Fig. 4.3 (a) Hold the pipette in one hand. Pipette-handling with forefinger over the mouth end to allow accurate control of delivery. Notice that the tube cap is held with the little finger of the same hand. (b) Dry lips are placed around the mouth end of the pipette while holding the forefinger of the right hand in ready and pipetting liquid in pipette. (c) Adding liquid into a slightly opened Petri dish.

7. Take the melted stab agar (48°C) and wipe the tube to remove any spar final moisture on the tube. Tilt the tube quite sharply to ensure that no solid lumps of agar remain in any part of the tube.
8. Remove the cap and flame the neck of the tube prior to pouring medium. Pour about 20 mL medium (at 48°C) carefully into the sterile Petri dish containing 1 mL inoculum (Fig. 4.2d) and as soon as adding medium close the Petri dish lid.
9. Mix thoroughly by rotating the plates in the route drawing 8, whereas it is lying on the table top (Fig. 4.2e). But do not splash any medium up over the sides.
10. Allow the plate to solidify the agar on the table top (within 10 min); then invert the plate to prevent condensation from the lid dropping onto the agar surface. When water falls onto agar surface, the microorganisms tend to spread and do not produce discrete colonies.
11. Incubate the inverted plates at 35°C for 24 h in the incubator.
12. After incubation, examine the plates for single isolated colonies (Fig. 4.2f). Sketch appearance of colonies and indicate color of the colonies on plates. Record your results.

4.2.2 Spread-plate technique

1. Before beginning the procedure, label the bottom of the Petri plates, the name of microorganism, dilution factor, spread-plate name, and your initials using waterproof marking pen.
2. For this inoculation, remove 1 mL pipette from box (Fig. 4.2a). Open box slightly and move the pipettes up to mouth of box by tilting. Remove lid and flame the mouth of box and pipettes. Take the pipette, flame the mouth of box again, and place the cap of box.
3. Make successive 10 fold dilutions of bacterial culture by transferring 1 mL of sample into 9 mL sterile diluent in each case (Fig. 4.1). Mix the sample in diluent by vertexing (Fig. 4.2b) or rotating through hand (Fig. 4.2c). Use new sterile pipette for each of dilutions (Fig. 4.3a and b).
4. Hold the 1 mL pipette with one hand and take the tube containing the diluent with other hand.
5. Remove the cap of tube and flame mouth of tube.
6. At the same time of dilution procedure, remove 1 mL diluted or non-diluted sample using a sterile pipette (Fig. 4.3).
7. Flame mouth of the tube, replace the cap and place tube to receptacle.
8. Transfer 1 mL into slightly opened sterile Petri plate in duplicate under aseptic conditions (Fig. 4.3c) and replace plate lid.
9. Place used pipettes into pipette discard pod.

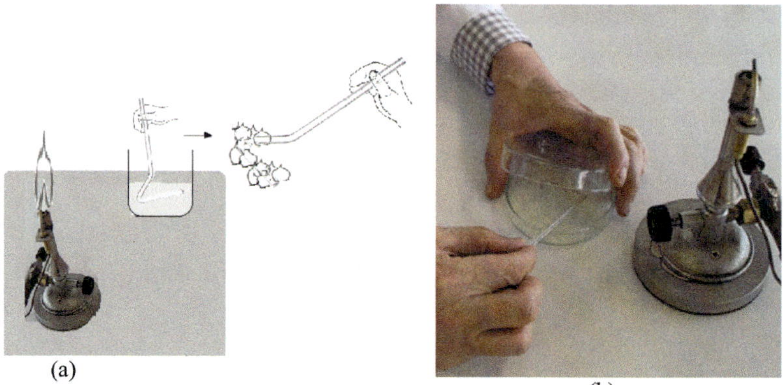

(a) (b)

Fig. 4.4 (a) Flame sterilizing an alcohol-dipped spreader. (b) To avoid killing the microorganisms on plate, cool spreader by touching it to the area in a part of the plate away from where added the culture. Spreading inoculum with the bent glass spreader while turning the plate.

10. Remove a glass spreader, soak into ethanol in a beaker, pass through flame and allow to burn the alcohol near the flame (Fig. 4.4a). Open lid of plate and cool the spreader by touching to the surface area of agar medium away from where added the sample (Fig. 4.4b).
11. Spread inoculum over the surface of the medium by pushing the glass spreader backward and forward while rotating the plates.
12. Replace plate lid and leave on tabletop to dry surface.
13. Incubate inverted plates at 35°C for 24 h in the incubator.
14. After incubation, examine the plates for single colony formation (Fig. 4.2f). Sketch appearance of colonies and indicate color of the colonies on plates. Record your results.

4.2.3 Streak-plate technique

1. Before beginning the procedure, label the bottom of the sterile Petri plate with the name of microorganism, dilution factor, streak-plate name, and your initials using water-proof marking pen.
2. Hold the tube containing the broth culture in one hand. Shake the broth culture gently to suspend the cells but avoid wetting the plug.
3. Take the loop by other hand. Flame the loop to redness and allow to cool for 4 s near flame.
4. Remove the cap of the tube using the little finger of the hand holding the loop and immediately flame the lip of the tube mouth.

5. Introduce the loop into the broth and withdraws one loopful of culture. Or remove an inoculum by touching loop onto center of selected colony (Fig. 4.5a-1). Do not dig into the agar with your loop during pick up microorganism and do not completely remove lid to expose surface to air.
6. Hold the Petri plate on the desk top or the bottom rests on the palm of the hand. Lift the lid of Petri plate and place the loop containing the culture onto the side of the agar surface and gently move loop on the surface area of agar to add the culture onto agar medium (Fig. 4.5a-2).
7. Lower the lid of the plate, flame the loop and cool loop for at least 4 s near flame and further cool the loop by touching onto the surface area of agar (Fig. 4.5a-3).
8. Streak (quadrant streak) the inoculum with four parallel lines on the agar surface from side to side starting from initial inoculum (Fig. 4.5b). Do not dig loop into the agar. These are the first streaks.
9. Lower the lid of the plate and flame the inoculating loop.
10. Rotate the Petri plate one quarter of a full turn slightly; raise the lid of plate and cool the loop by touching the agar surface away from the

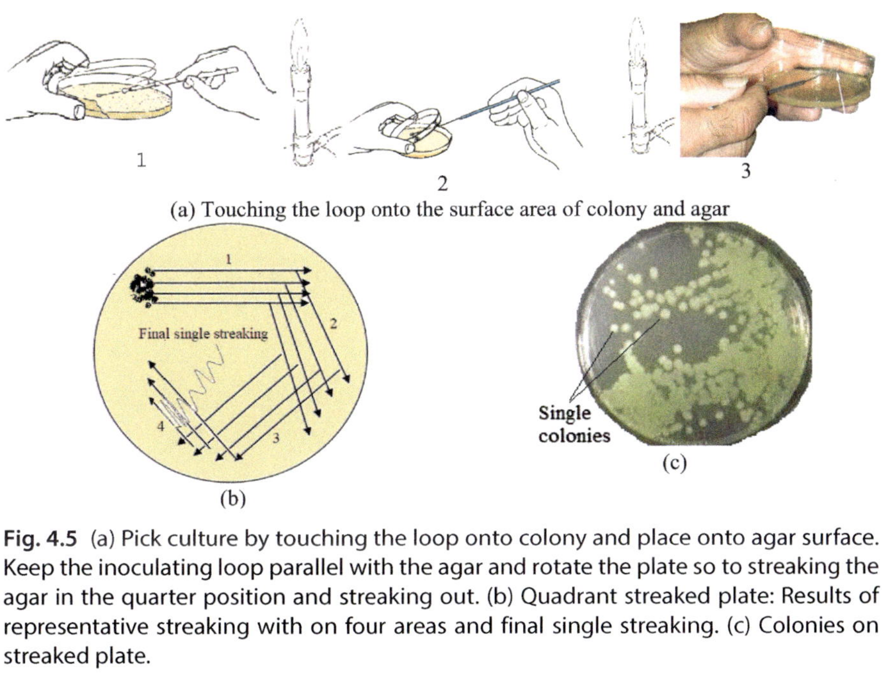

Fig. 4.5 (a) Pick culture by touching the loop onto colony and place onto agar surface. Keep the inoculating loop parallel with the agar and rotate the plate so to streaking the agar in the quarter position and streaking out. (b) Quadrant streaked plate: Results of representative streaking with on four areas and final single streaking. (c) Colonies on streaked plate.

set of streaks. Draw four lines starting from each first streaked line and make a second set of four streaks (Fig. 4.5b).

11. Follow steps 9 and 10 again for third and fourth streaking (Fig. 4.5b).

12. Without sterilizing loop after fourth streaking, a final single streaking should be made starting from fourth streaks (Fig. 4.5b). Sterilize the loop by flaming before putting it to the receptacle.

13. Incubate the inverted plate at 35°C for 24 h in an incubator.

14. After incubation, examine the plates for single colony formation. The possibility of single colony formation is high on final single streak (Fig. 4.5c). It will be observed that away from the initial streak, the growth is less dense and discrete colonies are present. Record colony features (color, shape, appearance, etc.).

LABORATORY REPORT

Date:
Name:

Lab Title: Pure Culture Techniques

1. **Observe isolated colonies on plates and record your observations for:**

   Streak plate:

   Spread plate:

   Pour plate:

2. **Explain how dilution is a common approach employed by any pure culture technique?**

3. **How would you obtain 1:100 dilution?**

4. **Why must plates be inverted during incubation?**

5. **Where and how should a label be written on an agar plate?**

6. **Before inoculating and pouring melted nutrient agar into a plate, why must the agar have cooled to 48°C?**

7. **Explain importance of plating techniques.**

8. **Why the use of aseptic technique is essential during working with pure microbial cultures?**

9. **A bacterial species is inoculated on EMB agar.**

   (a) The bacteria do not grow. Why?

   (b) If the bacteria ferment lactose, what would you expect to see?

   (c) The bacteria produce clear colonies. Why?

10. **What is colony-forming unit (cfu)?**

PRACTICE 5

Culture preservation techniques

5.1 Theoretical information

Once a microorganism has been isolated and propagated as a pure culture, the culture can be stored for a longer time period as a pure culture with the protection from contamination. During storage in laboratories, the viability of microbial cells in cultures is retained by transferring periodically onto (or into) a fresh medium (sub-culturing). Transfer is always done under aseptic conditions to avoid contamination. The working stock culture (short-term maintenance) will be used in the laboratory experiments or routine studies. Too old cultures may be damaged in some way; it should be changed with fresh culture. Bacteria and fungi for class use are generally preserved on nutrient agar and potato dextrose agar slants or deeps respectively.

Molds grow in refrigerator at about 5°C. Mold spores can easily germinate in the presence of lower amount of moisture on cotton stopper of tube. The mycelia may penetrate the cotton stoppers and contaminate cultures. For this reason, it is better to use screw-cap tubes for stock cultures. Such tubes also prevent drying of the cultures. Some microorganisms require addition of specific nutrients, such as serum, blood, and other special constituents; anaerobic microorganisms, which require the absence or marked reduction in the amount of free oxygen, and certain soil microorganisms require special inorganics. During preservation of culture, the toxic metabolic waste products are released by microorganisms diffuse into porous agar away from microorganisms. Therefore, microorganisms will die unless transferred to fresh medium at frequent intervals.

Repeated subculture is time consuming. It is difficult to successfully store many pure cultures for a long time. There is a risk of genetic alteration as well as contamination and drying during storage. Stock cultures on slants stored in the refrigerator at 5°C should be transferred to fresh media about once every 4 months. They can be stored for many years by lyophilization (freeze-drying), in liquid nitrogen at −196°C and in freezers from −70 to −120°C.

Storage conditions (such as temperature, oxygen, and humidity) affect the viability of microorganisms. Drying and freeze-drying can cause changes in different characteristics of microorganisms. These effects on the microbial cells can include changes on colonial morphology and pathogenic ability.

Laboratory Practices in Microbiology
https://doi.org/10.1016/B978-0-323-91017-0.00006-8

The factors in the recovery of freeze-dried microorganisms important such as such as the length of time allowed for re-hydration and whether additives added to aid recovery. Most samples can be left for 30 min post rehydration of fungi in 0.1% peptone water. For bacteria, recovery is best achieved using broth suspensions or growth media. It is generally accepted that rapid thawing gives best recovery of microorganisms. Rapid thawing avoids re-crystallization of ice which could cause mechanical damage. A warming can be achieved in a water bath at 37°C. Recovery microorganisms should be inoculated onto an appropriate nutrient source and stored in optimal conditions.

5.2 Preservation techniques

Materials

Stock culture of microorganisms, broth media, solid media, sterile glycerol (or 5% dimethyl sulfoxide), 10% skim milk, 70% ethanol, water bath, and sterile vials.

5.2.1 Testing for purity of preserved culture

Stock culture is inoculated into suitable nutrient broth and incubated at optimum growth temperature. After incubation, a loopful of broth culture is streak plated onto a suitable nutrient agar. When the colony is transferred from the plate, an inoculating needle (straight wire) is used instead of the wire loop. The needle is touched onto the center of the colony where there is a greater probability of getting only one species of microorganism. A staining from broth culture can be made to check purity of culture.

1. Take the loop by one hand and flame the needle to redness. Cool for at least 4 s near flame.
2. Stock culture is inoculated into suitable nutrient broth using cooled loop and incubated at optimum temperature until growth.
3. Take the loop by one hand and flame the needle to redness. Cool for at least 4 s near flame.
4. Incubated broth culture is streak plated onto suitable agar plate with cooled loop and incubated at optimum growth temperature.
5. Select a well-isolated colony from plate culture and label the selected colonies by numbering.
6. Label nutrient broth tube with selected colony number and name of microorganism.

7. Take the needle by one hand and flame the needle to redness. Cool for at least 4 s near flame.
8. Lift the Petri plate lid top and touching needle onto a clear agar surface for further cool.
9. Remove an inoculum by touching needle onto the center of the selected colony. Close the lid of plate. Do not dig needle into the agar medium and do not completely remove the lid of the plate exposing the surface of agar medium to the air.
10. Take a sterile broth in the tube and remove the cap from the tube by grasping with little finger of the hand holding the needle. Flame the mouth of tube.
11. Insert the needle into the broth, inoculate microorganism by dispersing the microorganism into the medium with moving the needle back and forth in the tube in a short height.
12. Flame the inoculating needle and place it on the receptacle.
13. Incubate tubes at optimum growth temperature of microorganism.
14. After incubation, look for growth.
15. Carry out streaking from broth culture onto the nutrient agar plate using loop and incubate inverted plates at optimum growth temperature of microorganism.
16. After the incubation, examine the plate for pure well isolated single colonies and purity of colonies with comparing color and appearances. Label the colonies with colored pens.
17. Apply Gram-staining on culture from petri plate, look at for purity of cells and other characteristics under microscope. Record results.
18. If culture is pure as indicated from colony morphology and microscopic analysis, carry out inoculation from selected single colonies onto agar slants and into agar stab.
19. Incubate the tubes at optimum growth temperature of microorganism.
20. After incubation, the cultures are stored as stock culture in refrigerator or freezer.

5.2.2 Short-term preservation techniques for microbial cultures

This method involves the inoculation of stock culture into fresh agar medium. Bacteria and fungi are generally preserved on nutrient agar and potato dextrose agar slants or deeps respectively for class or short-term use.

1. Label the media for the name of microorganism, name of media, date and your name or initials.

2. Carry out inoculations from stock culture into suitable broth media using loop.
3. Incubate the broth at optimum growth temperature of microorganism.
4. After incubation, remove a loopful of broth culture and inoculate each slant with a straight streaking from the bottom to the top up to half distance of agar slant. A straight streak is more useful than one that is spread over the entire surface.
5. With a needle, inoculate the agar deep from broth culture by inserting needle into the deep up to half distance of agar.
6. Incubate the tubes at optimum growth temperature of microorganism.
7. After incubation, examine the slants and deeps to note the extent of growth. Some microorganisms require close examination to see the growth, especially if the growth is thin and translucent.
8. Store the culture in refrigerator or in deep freezer for future use.
9. Remember following points concerning your slant or agar stab stock cultures:
   - Most stock cultures will keep for 4 weeks in the refrigerator. Some fastidious pathogens will survive for only a few days.
   - Do not use stored stock culture for making slides or routine inoculations.
   - Cultures begin to deteriorate after 2 or 3 days at room temperature. Some cells die out completely.
   - When frozen culture is melted due faulty in storage condition, does not freeze culture again and you must perform serial transfer.

Pure cultures storage on an agar slant or stab medium either in refrigerator (at 0–4°C) or in cold-rooms is applied for short duration (2–3 weeks for bacteria and 3–4 months for fungi) because the metabolic activities of the microbial cells will be very slow, not stopped and only less quantity of nutrients will be utilized. Thus, their growth continues slowly, nutrients are utilized and waste products released in medium. This results in the death of the microbial cells after sometime. This method cannot be used for a very long time because toxic products get accumulated which can kill the microbes.

5.2.3 Long-term preservation techniques for microbial cultures

Microbial cultures can be preserved long term in the presence of cryopreservation or by freeze drying. Freeze-drying method is mostly preferred for preserving microbial cultures in microbial culture centers.

5.2.3.1 Storage at freezing

1. Inoculate pure culture onto an agar slant and agar deep from a broth culture.
2. Incubate at optimum growth temperature of microorganism.

3. After suitable growth, culture is covered with sterile glycerol and is stored in freezer with a temperature −20°C.
4. Some bacterial species can be stored from 6 months to 2 years by this technique.

5.2.3.2 Freeze-drying

This is a widely used long storage technique.

1. Prepare a 20% (w/v) solution of skim milk and sterilize in small volumes (5 mL) at 116°C for 20 min. Avoid overheating; it can cause caramelization of milk.
2. Sterilize vials containing filter paper at 180°C for 1 h.
3. Inoculate pure culture into nutrient broth using loop and incubate at optimum growth temperature of microorganism.
4. After incubation, harvest broth cultures (biomass) by aseptic centrifugation at 10,000 × g at 4°C for 30 min.
5. Suspend biomass in a tube containing sterile 10% skim milk (as cryoprotective agent) to obtain cell suspension containing at least 10^6 cells per ml.
6. The cell suspension is dispensed (0.2 mL) into vials.
7. Insert the cotton plugs about 1.3 cm below the rim of the vial.
8. Freeze cells in vials for 1 h in freezer at −65°C and dry culture under vacuum pump within 18 h.
9. At the end of drying, place in a dry box and allow cooling.
10. Seal the mouth of vial by heating just above cotton plug and allow vial to cool.
11. Freeze dried cultures are stored below 5°C. Room temperature storage of freeze-dried cultures should be avoided.

Many factors can affect the stability of the freeze-dried culture: the type of growth media, the age of culture, the growth phase of culture, and the type of microorganisms. The freeze-dried culture viability can be retained up to 20 years.

Rehydration. When culture is required, cut vial below 2.5 cm from the tip. The vial is disinfected with 70% alcohol and break at scored area. The opened end of vial is flame gently. Vials containing pathogenic bacteria should be opened only in a closed safety cabin. Rehydrate freeze-dried cultures immediately after opening by adding 0.2–0.4 mL of a sterile broth into vial. The vial is mixed well to dissolve pellet completely. Culture will rehydrate within a few min and transfer one loopful of rehydrated culture to a test tube containing 5 mL of same broth used in rehydration.

5.2.3.3 Ultra-freezing

1. Inoculate pure culture into nutrient broth and incubate at optimum growth temperature of microorganism.

2. Harvest cells (biomass) from broth cultures by aseptic centrifugation at 10,000 × g at 4°C for 30 min.
3. Suspend biomass with sterile fresh medium containing either 10% glycerol or 5% dimethyl sulfoxide.
4. Transfer 0.4 mL of the cell suspension containing at least 10^2 cells per ml into sterile vials.
5. Place the vials into freezing chamber and freeze cells at a controlled rate of 10°C per min to −30°C then at a more rapid drop of 15–30°C per min to −150°C.
6. After freezing, store vials in liquid nitrogen phase at −196°C or above.

Thawing. When the culture is required, rapidly thaw the culture with moderate agitation in a 35°C water bath until all ice melts. This usually takes 40–60 s for glass vial. Immediately after thawing, wipe vial mouth with 70% ethanol to disinfect. Open vial and aseptically transfer culture to fresh medium.

5.2.3.4 Immersing in paraffin (or mineral oil)

Storage of microbial slat culture in the sterile liquid paraffin is an economical method.

1. Liquid paraffin (or glycerol) is sterilized in an oven at 180°C for 1 h.
2. An agar slant is inoculated with pure culture from a broth culture and incubates at optimum growth temperature of microorganism.
3. After suitable growth, aseptically sufficient sterile paraffin is added into tube to cover slant.
4. Store oil covered cultures in an upright position at refrigerator temperature (or frozen state at −70°C). Culture under the sterile paraffin can be stored from months to years.

Covering of culture with paraffin will prevent dehydration, reduce metabolic activity, reduce microbial growth (anaerobic), and allow microorganisms in dormant state.

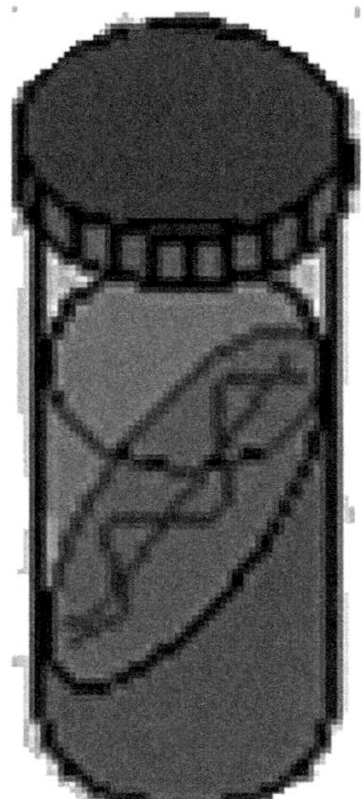

In paraffin

When culture use is needed. A small amount of growth (colony) under oil can be transferred with an inoculating needle to fresh broth. Take care when flaming needle since oil will splatter and contaminate surrounding area and personnel.

5.2.4 Novel culture preservation techniques

Novel techniques on culture preservation have emerged as an alternative technique to the conventional culture preservation methods. Novel techniques are simple and effective to improve cellular properties while preserving the cellular functionality during processing and storage.

Cellular immobilization retains the stability of the microorganisms during processing and storage. Unlike traditional storage methods (such as freezing and freeze drying), which can lead to irreversible protein denaturation, cell viability, and membrane damage, microencapsulation increases the stability of microorganism cells during storage and becomes cells. The effectiveness of the encapsulation technique depends on the type and concentration of the capture material used and the application form of the method. In addition, parameters, such as the size of the material used, its pore structure, texture, and the nature of the coating material can support cell protection and increase the efficiency of the encapsulation process. In addition, when microbial cells are stored at $-80°C$ without the use of cryoprotectants, microencapsulation can have a protective effect on viability in probiotic strains.

SECTION II

Techniques on morphological examination of bacteria

Bacteria are classified according to their morphological appearance by viewing under a light microscope. The shapes of bacteria are coccus, bacillus, and spirals. But there may be other shapes such as helical bacteria (spirochetes), curved bacteria, and unusual shaped bacteria. There are many microscopic techniques in the evaluation of morphological characteristics of microorganisms. Microscope is an important tool for the laboratory as the microorganisms are invisible to the naked eye unless magnified. The light microscope has two or more lens systems. The arrangement of lenses increases the magnification of the object many folds depending on the magnification of eyepiece and objectives lenses. Microscopy is used in laboratories for many purposes, including diagnosis of infectious diseases, identification of microorganisms in environmental samples (e.g., food and water) and others. The microscopic techniques in this section will familiarize you with the microscopes to look at various types of microorganisms and identify their different morphological structures.

The practices used in the morphological examination of microorganisms in this section are as follows:

Practice 6. Microscope for Study of Microorganisms
Practice 7. Microbial Cell Measurement Technique
Practice 8. Wet-Mount and Hanging-Drop Techniques
Practice 9. Smear Preparation and Simple Staining Techniques
Practice 10. Gram Staining Technique
Practice 11. Bacterial Endospore Staining Techniques
Practice 12. Bacterial Capsule Staining Techniques
Practice 13. Motility Tests

PRACTICE 6

Microscope for study of microorganisms

6.1 Light microscope

A binocular microscope is illustrated in Fig. 6.1 and functions of its parts are given in Table 6.1.

Framework. The arm and base of a microscope are basic frame structures in all microscopes. All other parts of the light microscope are attached to the frame of the microscope.

Stage. The horizontal platform supporting the slide of the light microscope is called a stage. Slide on the stage is moved with a mechanical stage.

Light source. In the microscope, the intensity of the light is changed by voltage control. Under the microscope, the amount of light that can be delivered to the microscope can be controlled with a wheel located near the base of the microscope. It has another button on the microscope base that controls the voltage. Most microscopes have a filter control mechanism to reduce light intensity.

Lens systems. Light microscope has three types of lens system in the magnification of microorganisms: oculars, objectives, and condenser. It has the ocular (or eyepiece) at the top and it consists of two or more oculars and usually has 10X magnification in microscopes. Light microscopes usually have three or more objectives. Objectives present on the rotatable head piece, which makes it possible to move them over a work area. The most common objectives on the light microscopes are called low power, high power, and oil immersion; they have 10X, 44X, and 100X magnifications, respectively. Sometimes a fourth objective with 4X magnifier can be attached to microscope for fast field scanning. The third lens system is the condenser located in the lower part of the microscope. The condenser collects the light and sends it to the slide of the sample to be analyzed. The condenser can be moved down and up with a knob under the stage. The amount of light reaching the slide is regulated by the diaphragm in the condenser.

Oil-immersion objective. Oil-immersion objective can magnify object about 100X. To obtain the best possible result, the objective is immersed in a medium that has approximately the same index of refraction of glass,

Laboratory Practices in Microbiology
https://doi.org/10.1016/B978-0-323-91017-0.00003-2

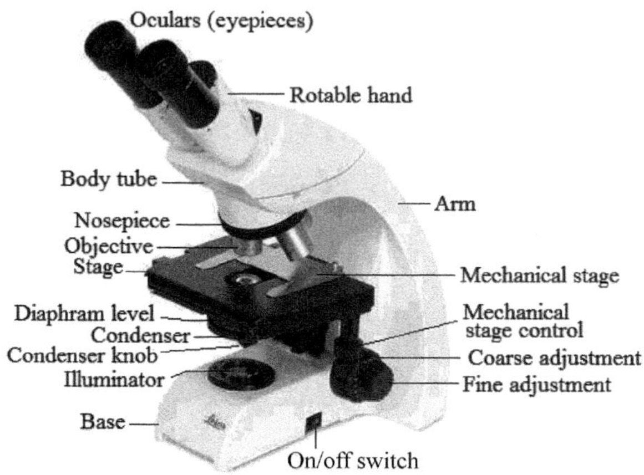

Fig. 6.1 Parts of a binocular microscope.

Table 6.1 Parts and their functions of a binocular microscope.

Part	Function
Base	Supports entire microscope.
Mechanical stage with slide holder	Slide holder allows placing the slide in the mechanical stage and allows the slide to move.
Body tube	Raising and lowering the objective during focusing.
Resolving nosepiece	Rotating to change from one objective to another.
Stage	Supports mechanical stage and microscope slide.
Diaphragm lever	The amount of light striking the object can be controlled with opening and closing by diaphragm lever.
Condenser	Condenses light waves into a pencil-shaped cone, prevents the escape of light waves, and controls light intensity when raised or lowered.
Condenser knob	Raises and lowers condenser.
Illuminator	Reflects light waves from light source into condenser.
Mechanical stage control	Move the slide on two horizontal planes, back and forth, and side to side.
Arm	Upper half parts of the light microscope are supported by arm
Coarse adjustment	Coarse adjustment is moved up and down to provide an approximate view.
Fine adjustment	The image is adjusted by moving the body tube up and down slowly using fine adjustment.
Nosepiece	Makes possibility of movement of objective into position over a slide.

1.6. One medium commonly employed for this purpose is immersion oil. Its index of refraction is the same as that of glasses; oil does not bend the light rays entering the front lens of the oil-immersion objective. If air rather than oil is present between object and objective, some light is lost.

Ocular adjustment. On a binocular microscope, the distance between two oculars can be changed according to the examiner eye distance. The distance between the slide and objective can be adjusted by the ocular adjustment.

Resolution. The resolution power of the light microscope lens system is a function of its wavelength of light, numerical aperture, and capacitor. The magnification of a microscope is adjusted by its resolving power or the ability to distinguish images of two nearby objects separately. Optimum resolution of the oil-immersion lenses on the light microscope is around 0.2 μm. With this resolution power, two microbial cells with 0.2 μm apart will be magnified as separate entities.

Perfocal. The objectives on the microscope are perfocal. When a microbial preparation is magnified with one objective, it will be in approximate magnify with the other objectives when they are rotated into working position. Thus, if you have an object in sharp magnification on low-power objective, you can rotate the high-power objective into the working position and achieve sharp magnification with only a slight turning of the fine adjustment knob.

Condenser. The substage condenser receives light from the microscope light source and concentrates it into a cone of light that illuminates the specimen with uniform intensity over the entire view field (Fig. 6.2). It is critical that the condenser light cone be properly adjusted to optimize the intensity and angle of light entering the objective front lens.

The condenser receives light from the light source of the microscope and concentrates to the sample through a light path (Fig. 6.2). The condenser light cone must be properly adjusted to optimize the intensity and angle of light entering the front lens of the lens.

Total magnification on microscope. Fig. 6.3 illustrates the preparation of light microscope for magnification of microorganisms. Microscope total magnification is equal to the multiplication of ocular magnification power with objective:

Ocular		Objective	Magnification at eyepoint
10X	X	10X	100X
10X	X	44X	440X
10X	X	100X	1000X

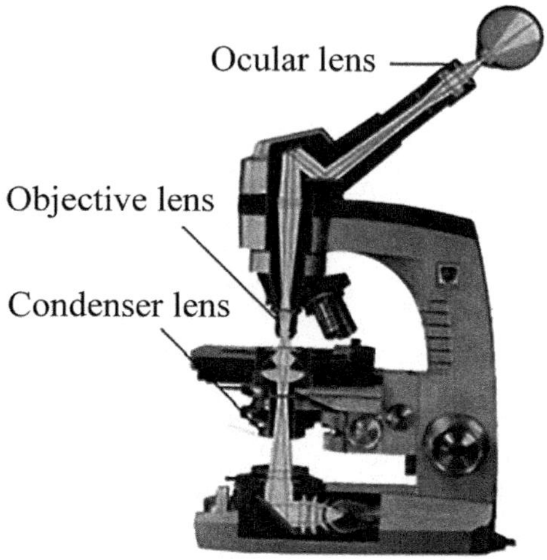

Fig. 6.2 Light pathway. *(Modified with permission from McGraw Hill; H.J. Benson, Microbiological applications: laboratory manual in general microbiology, 2001, Fig. 1.3, p. 4.)*

Care of the microscope

If certain precautions are not taken for microscopes in the laboratory, they can easily be damaged. Microscope and lens systems can be kept clean with the following suggestions:

1. Do not touch the microscope lens in any way. If the lenses become dirty, it can be cleaned by lightly wiping them with lens paper.
2. After your work is completed, never leave slides on the microscope.
3. If the oil is used in the analysis of the samples, the oil on the objective must be cleaned. If the oil dries and hardens on the lens, the microscope lens will deteriorate, so the oil on the lens must be slightly moistened with xylol and cleaned after the work is completed.
4. If any liquid drops or spills on the microscope present, clean it with a piece of clean cloth and dry it. If the immersion oil remains on the stage of the microscope, wipe it after moistening with xylol using a piece of paper, then clean and dry the stage areas.
5. If there is immersion oil in the microscope, do not bend the microscope.
6. When the microscope is not used, keep the microscope closed and replace to the microscope room.
7. **Transport.** When moving the microscope from one part of the laboratory to another, carry it with both hands (Fig. 6.4).
8. **Solvents.** Various fluids can be used in the cleaning of microscope lenses. The use of warm water and green soap is the best cleaning agent. Xylene is one of the other universally accepted solvents. Alcohol and

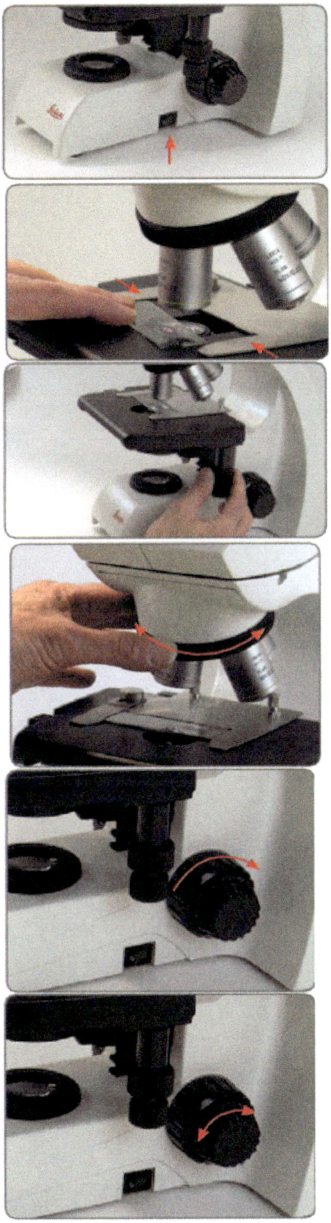

Turn on microscope switch on the bottom right-hand side of base.

Place the slide containing sample on the stage by sliding slide forward into the slide grips (slide grips hold the slide in place).

Use stage control to position the slide so a portion of the sample is under the objective in use.

Rotate the nosepiece (using the knurled nosepiece ring) to move the low power objective (10x) into working position.

Raise the stage by rotating the coarse adjustment knob to its positive stop at the highest position. Find microbial cells while looking only through one of eyepieces. To help focus, cover or close another eye.

Using the fine adjustment knob, bring the sample into sharp focus so that it is comfortable for viewing while looking only through one of the eyepieces.
Switch to a high power objective (not an oil objective) and focus the microscope while viewing with both eyes.

Fig. 6.3 Preparation of light microscope for magnification.

(Continued)

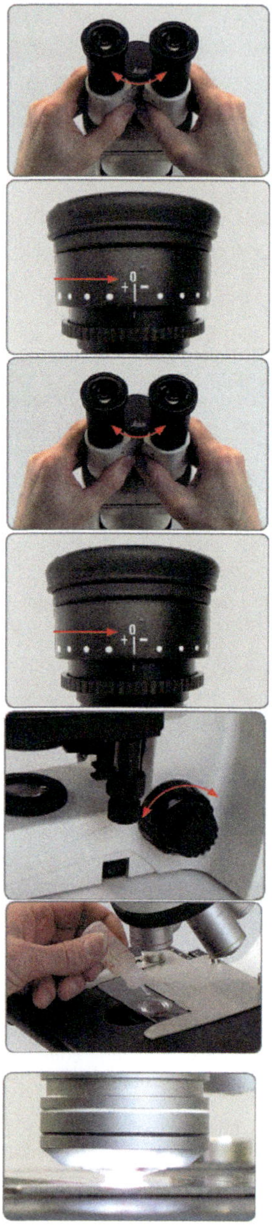

Adjust eyetubes for your interpapillary distance. Fold or unfold eyetubes to decrease or increase the distance between eyepieces until you see one illuminated circle.

Set focusing eyepieces to "0" diopter setting.

Adjust eyetubes for your interpupillary distance. Fold or unfold eyetubes to decrease or increase distance between eyepieces until you see one illuminated circle.

Set focusing eyepieces to "0" diopter setting.

Swing low- or high-power immersion objective and focus the microscope while viewing with both eyes.
Find the field of view on the slide that you would like to inspect using coarse snd fine knobs.

Switch the oil-immersion objective one half. Place a drop of immersion oil onto preparation on the slide.
Swing the oil-immersion objective in place.

Look through the microscope and slowly rotate the fine focus knob to raise the stage until the sample is in focus.
When finished using the oil-immersion objective, be sure to clean the front of the objective and any other surfaces that the oil contacted.

Fig. 6.3, cont'd

(a) (b)

Fig. 6.4 Carrying microscope. (a) Wrap the cord. (b) Carry microscope by handle on back of microscope and an undercut in front.

acetone can also be used as solvents. Acetone is a powerful solvent for oils that can be found on objective lenses.

6.2 Uses of microscope

Following suggestions should be considered for setting up a microscope and performing the microscopic magnification.

6.2.1 Examination with low-power objective

The main reason for starting in the magnification of object with the low-power objective is to enable to look for the object on the slide and prevent damaging of the objectives. When you find what you are looking for a general finding of objects in the objectives, you can switch to the low magnifier objective. During performing analysis of object on slide under a light microscope using low-power objective, the following steps should be used:

1. Place the microscope slide containing microbial preparation on the stage. Fig. 6.5 illustrates the place of slide on the light microscope for magnification.
2. The light source is turned on to use a minimum amount of voltage in the microscope. If necessary, the slide can be repositioned in the middle of the light source.
3. It should be checked that the condenser is raised to its highest point.

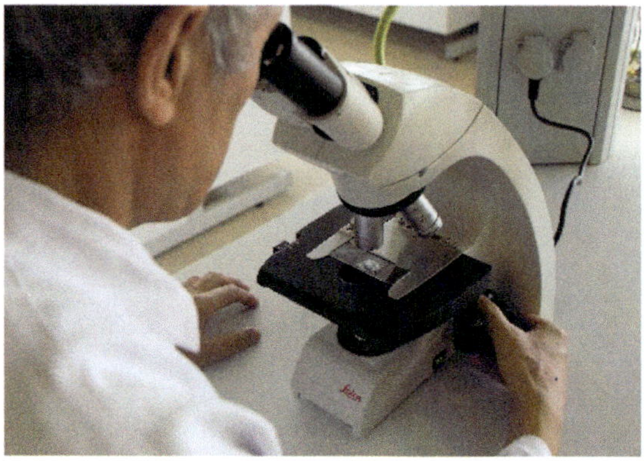

Fig. 6.5 Microscope position for viewing.

4. If the low-power objective is not above the area to be analyzed, rotate the objective to be analyzed. Make sure that the locked position is unlocked when rotating the target.
5. The coarse adjustment knob is turned until you find the object under the microscope. A built-in stop will prevent the objective from touching the slide.
6. When looking at the image from the eyepieces (or oculars), the coarse or fine adjustment is rotated to focus the object. After finding microbial cells in the preparation, focus on a better illustration using the fine adjustment knob. Don't readjust the coarse adjustment knob.
7. To achieve the clearest image, reducing or increasing the light intensity can be achieved using the aperture lever. The diaphragm should be closed carefully while reducing the light intensity.
8. A slide is moved with a mechanical actuator so that an image can be found and sought.
9. Once you have identified the object under low-power objective and wish to increase the magnification, you can proceed to either high-power or oil-immersion magnification.

6.2.2 Examination with high-power objective

The main reason for starting in the magnification of object with the low-power objective is to enable to look for the object on the slide and to prevent damaging of the objectives. To switch the low magnifying lens to the high magnifying lens, it is necessary to return the high magnifying lens to its working position. Adjustment with the fine adjustment knob may be required to sharpen the image under the microscope, but the coarse adjustment button should not be touched. If a microscope lens is perfocal, only minor focus adjustments are needed when switching from a low magnifying lens to a high magnifying lens because all the objectives will be parfocalized. Nonperfocalized microscopes require considerable refocusing when changing objectives. The high-power objectives should be used only together with cover glasses; without them, images are usually unclear.

6.2.3 Examination with oil-immersion objective

If an immersion oily lens is to be used in the imaging, special immersion oil is added between the lens and the microscope slide. The immersion oil is used because it has the same refractive index as the glass slide, which prevents light loss due to the bending of light rays while passing through the air. The use of immersion oil increases the imaging power of the microscope.

In perfocalized microscopes, oil-immersion objective can be passed from low-power or high-power objective without image loss. In microscopic imaging, it is best to switch from low-power to high-power objectives and from high-power to oil-immersion objective. When the microscope is focused at a magnification using high power objectives, the oil-immersion target can be returned to its position without fear of hitting the slide. A drop of immersion oil should be placed on the slide before returning the oil-immersion objective to the position. The oil-immersion objective should never be used without oil. By the way, if the oil looks cloudy, it should be discarded.

LABORATORY REPORT

Date:

Name:

Lab Title: Microscope for Study of Microorganisms

1. **List three fluids that may be used for cleaning the lenses.**

2. **What characteristic of a microscope enables one to switch from one objective to another without altering the focus?**

3. **What is the magnification of objects observed through an oil-immersion objective (100X) with an eyepiece (7.5X)?**

PRACTICE 7

Microbial cell measurement technique

7.1 Theoretical information

7.1.1 Bacterial shapes and arrangements

Bacteria are unicellular prokaryotic microorganisms. They divide by binary division (fission). In this division, one bacterium cell divides to two cells, two cells to four cells, four eight cells, and so on. There is a geometric increase in the population of bacteria at each division time (generation time). Many bacterial species have characteristic shapes and arrangements that are useful in classification and identification. There are three common shapes of bacteria: coccus, bacillus, and spiral.

7.1.1.1 Coccus (spherical) bacterial shapes

Overall shape of coccus (plural cocci) is usually spherical. Some of the spherical cell appears oval, elongated, or slightly flattened on one side. The average diameter of most bacillus is 0.5–2.0 μm (μm) in diameter and may be seen based on their planes of division and tendency to attachment of daughter cells after replication. They have one of several distinct arrangements based on their planes of division. One of the following arrangements appears after cell division (Fig. 7.1a):

(i) Micrococcus. Cells are spherical in shape. Cell division occurs in one plane, daughter cells separate from each other and they appear single, this form is called micrococcus. For example, *Micrococcus melitensis*.

(ii) Diplococcus. Cells are spherical in shape. Cell division occurs in one plane, daughter cells attach to each other and they appear in pairs. This arrangement of cells is called diplococcus. For example, *Neisseria meningitidis*.

(iii) Streptococcus. A bacterium with spherical cells arranges in chains. Cell divides in one plane and cells attach to each other to form longer chains with four or more cocci. This cell arrangement is called streptococcus (coccus in chains). For example, *Streptococcus thermophilus*.

Laboratory Practices in Microbiology
https://doi.org/10.1016/B978-0-323-91017-0.00023-8

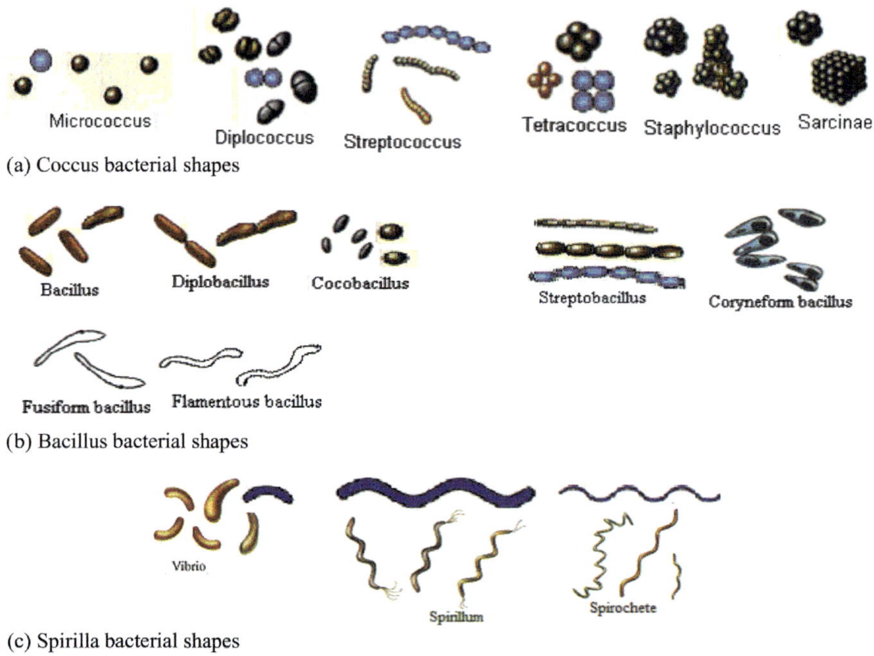

Fig. 7.1 Spherical, bacillus, and spiral shape of bacteria.

(iv) Tetracoccus. Spherical bacterial cells divide in two planes and daughter cells arrange in a group of four. This appearance is called tetracoccus. A tetrad (four cells) arrangement looks almost like a square under microscope. For example, *Gaffkya tetragena* and *Micrococcus luteus*.

(v) Staphylococcus. Coccus divides in a random plane, daughter cells attach to each other and irregular groups occur much like in clusters, grape-like shapes. This arrangement of cell is called staphylococcus. For example, *Staphylococcus aureus.*

(vi) Sarcinae. Spherical bacterial cells divide in three planes and cells arrange in a group of eight with forming a small cube. This arrangement of cell is called sarcinae (a cube of eight). For example, *Sarcina mirabilis*.

7.1.1.2 Bacillus (rod) bacterial shapes

The average length of rod-shaped bacteria is 1.0–10 μm in length and 0.25–1.0 μm in diameter. Rod-shaped bacteria (cylindrical) or bacillus always divide in one plane and there are subtypes according to the cell shape and arrangement of cell after division. Rod-shaped bacterium has one of the following shapes and arrangements (Fig. 7.1b):

(i) Single bacillus (single, bacilli): Most bacillus appears as single rod after cell division. Cylindrical shape has a diameter 1 μm and length 2–5 μm and is mostly single cell, but sometimes form pairs and chains. For example, *Escherichia coli* and *Salmonella* Typhimurium.

(ii) Diplobacillus. A rod-shaped bacterial cell divides in one plane; the cells cannot separate from each other after division. The cells appear pair. This type of arrangement is called diplobacillus. For example, *Coxiella burnetii* and *Klebsiella pneumoniae.*

(iii) Cocobacillus. Bacillus like coccus, very short, stumpy, oval, and usually short that they appear ovoid. For example, *Yersinia enterocolitica* and *Haemophilus influenzae.*

(iv) Streptobacillus. The rod-shaped bacterial cells divide in one plane and are arranged in chains. A bacterium with rod-shaped cell arranges in end-to-end chains. Streptobacillus often resembles to link of sausages. This type of arrangement is called streptobacillus. For example, *Bacillus subtilis* and *Streptobacillus moniliformis.*

(v) Coryneform bacillus. Bacillus bends at the point of division following the cell division, resulting in a picket fence and angular patterns that look like Chinese letters. They are irregularly rod-shaped bacteria, arrange at angles to form different arrangements, such as V- and L-shaped. For example, *Corynebacterium diphtheriae.*

(vi) Fusiform bacillus. When the bacillus is slightly curved at both ends after cell division, this is called fusiform bacilli. For example, *Fusibacterium necrophorum.*

(vii) Filamentous bacillus. When the bacillus grows in long threads (thinness) with turns, it is called filamentous bacillus. For example, *Streptomyces vibriviridochromogens.*

7.1.1.3 Spirilla bacterial shapes

Spirilla (sing. spirillum) shapes are curved-shaped bacteria which can range from a gently curved shape to a corkscrew spiral. Many spirilla are rigid and able to move. The length of rod-shaped bacteria is over 2–100 μm. Spiral-shaped bacteria occur in one of three forms: vibrio, spirillum, and spirochete (Fig. 7.1c):

(i) Vibrio. Vibrio are slightly curved or comma-shaped bacterial cells with less than one complete turn or twist in the cell, generally single and 2–3 μm long. For example, *Vibrio parahaemolyticus* and *Vibrio cholerae.*

(ii) Spirillum (plural, spirilla). A bacterium with rigid spiral (helical) structure (not easily band, not flexible), thick, long, and move with flagella, 6–15 μm long and spiral in shape is called spirillum. For example, *Campylobacter jejuni* and *Helicobacter pylori*.

(iii) Spirochete. It has a helical structure and flexible body (not rigid), a thin structure, long, spiral helical-shaped cells, or corkscrew-shaped spiral bacilli. Spirochete moves by axial filaments. For example, *Treponema pallidum* and *Leptospira borgpetersenii*.

7.1.1.4 Irregular bacterial shapes

Some of the irregular bacterial shapes are given in Fig. 7.2.

(a) Filamentous bacteria. They are very long thin filament-shaped bacteria. Some of them form branching filaments resulting in a network of filaments called "mycelium". For example, *Candidatus savagella*.

(b) Star-shaped bacteria. Fewer star-shaped cells are unusual shaped bacteria and have cells whose cross sections look like stars. Cell division is nearly symmetrical and begins with a slight elongation in two directions. The cross-wall separates the daughter cells, whose shape is flat, six-pronged stars, nearly triangular appendages, and lens shape. For example, *Stella humosa* and *Stella vacuolata*.

(c) Rectangular-shaped bacteria. They are box or rectangular-shaped bacteria (arcula). They are extremely halophilic and grow at temperatures between 40°C and 45°C. They are not pathogenic. For example, *Haloarcula vallismortis* and *Haloarcula marismortui*.

(d) Pleomorphic bacteria. These bacteria do not have any characteristic shape unlike all other bacterial shapes. This term refers to the ability to change or alter the size and shape under the pressure stress environmental factors. They can change their shape and have cilia. For example, *Mycoplasma pneumoniae* and *Mycoplasma genitalium*.

7.1.2 Measurement of microbial cells

The important physical properties used in the identification and classification of microorganisms are their sizes and shapes. The exact size of a

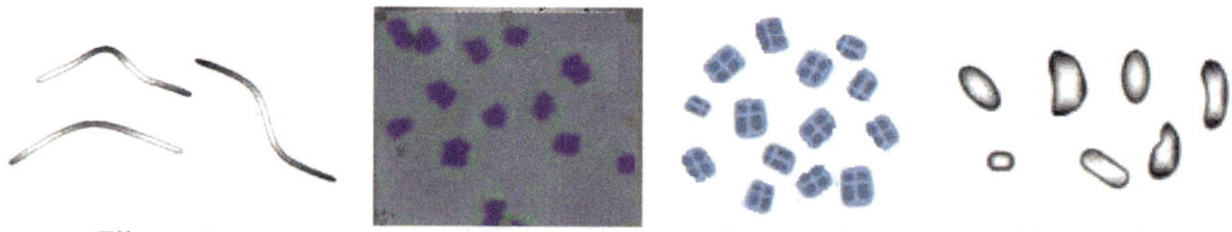

Fig. 7.2 Irregular bacterial shapes.

microbial cell in a microscope can be determined with a calibrated ocular micrometer for an objective of microscope. There are two types of micrometer: ocular and stage. The ocular micrometer is a circular glass disc with 100 equally spaced divisions, marked from 0 to 100. Calibration of the ocular micrometer is necessary because the distance between the ocular micrometer degrees varies with the size of the area and the target used. Ocular micrometer is calibrated with a known scale, stage micrometer. The stage micrometer is a glass slide with 1 mm divided into 100 equal parts. Each small division of the stage micrometer equals 0.01 mm with 10 large divisions and each large division is divided into small 10 divisions. Therefore, the value of the one division is 1/100 = 0.01 mm or 10 μm. The value of each division of stage micrometer increases with increasing magnification. To calibrate the ocular micrometer, the ocular and stage micrometers are imposed to microscope and the number of ocular lines is determined for each line of the stage micrometer. The ocular micrometer factor is then calculated for an objective. To determine the size of a microbial cell, the number of lines covered by the cell in the ocular micrometer is determined and multiplied by the ocular micrometer factor. Very rarely, a complete overlap can be achieved between the two micrometer lines.

Microbial morphological characteristics can be detected using microscope in the film prepared from culture. In the detection of size of microbial cell, there are three distinct steps in microscopic technique:

(a) Determination of the microscopic factor (MF),
(b) Preparation of the film from culture, and
(c) Examination of the film.

Once the MF of a microscope objective has been determined, the same MF can be used in future for respective object of microscope. The film is prepared by placing culture on a glass slide followed by spreading it over an area of 1 cm^2. The film is then dried and the characteristics of microorganisms are indicated. Micrometer measures the dimensions of microbial cells in terms of length, diameter, and thickness.

7.2 Cell measurement technique

Materials

Microbial cultures: *Staphylococcus aureus*, *Escherichia coli*, *Saccharomyces cerevisiae*, *Zygosaccharomyces baili*, *Aspergillus niger*, and *Penicillium roqueforti*.

Equipments: Ocular micrometer, stage micrometer, microscope slide in bottle containing alcohol and microscope.

7.2.1 Calibration of ocular micrometer for an objective

Scales on the ocular micrometer are calibrated to the standard grades of the stage micrometer for the objective. The ocular micrometer should be attached to the eyepiece.

1. Remove top cover of eyepiece of microscope (Fig. 7.3a).
2. Then, place the ocular micrometer carefully on the eyepiece. The eyepiece is placed back to the microscope and screwed to the top cap position.
3. After the ocular micrometer is installed into the eyepiece, low power objective image is taken to magnification position.
4. Adjust the light so that two different scales of ocular and stage micrometers are in sharp focus.

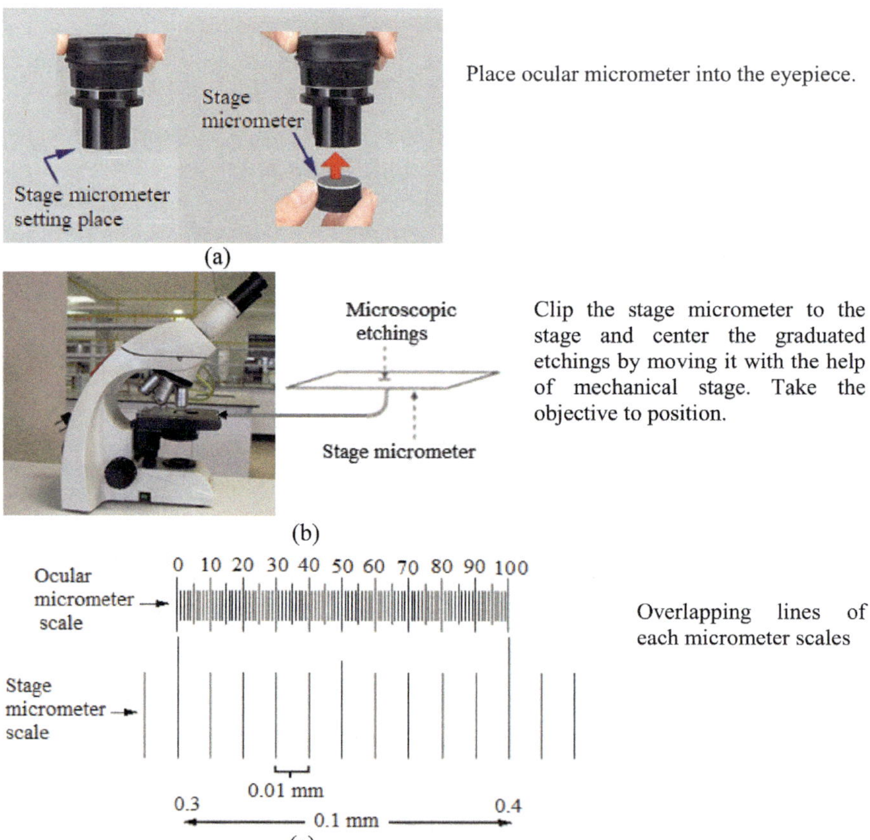

Fig. 7.3 Ocular micrometer calibration by stage micrometer for a microscope objective.

5. The stage micrometer is attached to the stage and it is centered on the objective image by moving the mechanical stage (Fig. 7.3b).
6. The eyepiece is rotated until the overlapping lines of each ocular and stage micrometer lines (Fig. 7.3c).
7. Focus the superimposed scales and coincide a line at left side and find another line set which coincides by focusing towards right side of the first coinciding line.
8. Count the division coinciding between two links in both micrometers. The number of ocular micrometer dimensions coinciding with the number of stage micrometer dimensions is found out (Fig. 7.3c).
9. Take at least seven readings randomly by following the steps 4–8.
10. Repeat the procedure with high power (44X) and oil-immersion (100X) objectives by following the steps 3–9.
11. Observations: Record your observations in the laboratory report and take the mean value of seven readings for low, high, and oil-immersion objectives.
12. From these results, the least count of the micrometer or the values of calibration factor of objective can be calculated in the following ways:

Scale of stage micrometer = 1 mm.
Since 100 divisions of stage micrometer = 100 mm.
1 division of stage micrometer = 1/100 = 0.01 mm or 10 μm.

(a) Low-power objective (10X) and eyepiece (10X)

Suppose 15 divisions of ocular micrometer = 25 divisions of stage micrometer.
1 division of ocular micrometer = 25/15 = 1.66 divisions of stage micrometer.
Since 1 division of stage micrometer 1.66 × 10 μm = 16.6 μm
Therefore, calibration factor can be calculated using following formula:
Calibration factor of low power objective = (Y/X) (10 μm) = (25/15) = 16.6 μm
X = Number of ocular divisions between the lines of coincidence
Y = Number of stage divisions between the lines of coincidence

(b) High-power objective (44X) and eyepiece (10X)

Suppose 16 divisions of ocular micrometer = 6 divisions of stage micrometer.
Calibration factor of high-power objective = (Y/X) (10 μm) = (6/16) = 3.7 μm

(c) Oil-immersion objective (100X) and eyepiece (10X)

Suppose 14 divisions of ocular micrometer = 2 divisions of stage micrometer.

Calibration factor of oil-immersion objective = (2/14) (10 μm) = (25/15) = 1.4 μm

7.2.2 Measurement of microbial cell

1. After calibration of ocular micrometer, the stage micrometer is removed from microscope stage.
2. Replace the stage micrometer by prepared microbial culture (cells/spores) on slide on the microscope stage.
3. The objective of microscope is taken to position.
4. Magnify the preparation with the objective (low, high, or oil-immersion so that the sells/spores are clearly seen.)
5. Count the number of ocular divisions covered by the microbial cell/spore by viewing through the eyepiece (Fig. 7.4). If cell/spore is not round, take length and width separately.
6. Measure the size of 10 cells or spores by moving the slide.
7. Write down your readings in a laboratory report and take the average.

Results: Determine the size of cell or spore using following formula:

Size of cell (or spore) = divisions of ocular micrometer covering by cell (or spore) X calibration factor of objective (μm)

Example: Suppose, on an average the cell occupies five divisions of ocular micrometer in high power objective (Fig. 7.4):

Size of cell = 5 X calibration factor of high-power objective (μm) = 5 × 3.7 = 18.5 μm.

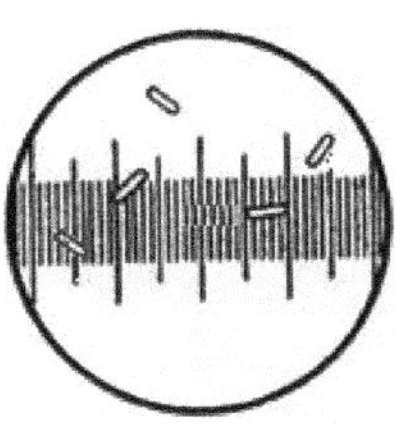

Measuring microbial cell size with the calibrated ocular micrometer. Here a cell occupies five divisions on the ocular micrometer

Fig. 7.4 Measurement of the size of microbial cells by calibrated ocular micrometer with the microscope objective.

LABORATORY REPORT

Date:
Name:

Lab Title: Microbial Cell Measurement Techniques

1. Draw the shape of the bacteria seen on the .... objective on each slide prepared with the bacteria cultures and calculate the size of the bacteria by counting the lines covered by the bacteria in the ocular micrometer.

Ocular micrometer readings: Stage micrometer readings:	Ocular micrometer readings: Stage micrometer readings:
Staphylococcus aureus	*Escherichia coli*
Shape =	Shape =
Arrangement =	Length (μm) =
Diameter (μm) =	Width (μm) =

2. Make drawings of yeast cells and calculates their size in .... objective.

Number ocular micrometer readings: Number stage micrometer readings:	Number ocular micrometer readings: Number stage micrometer readings:
Saccharomyces cerevisiae	*Zygosaccaharomyces baili*
Diameter (μm) =	Diameter (μm) =

3. **Make drawing of molds hyphae and their spores, and calculates their appearance under ………… objective.**

Ocular micrometer readings: Stage micrometer readings:	Ocular micrometer readings: Stage micrometer readings:
Aspergillus niger	*Penicillium roqueforti*
Hyphae type = Type of Spores =	Hyphae type = Type of Spores =

4. **What is a microbe?**

   What are some reasons why microbes are important?

5. **What would be the length of a microbial cell, to two significant figures, that was 35 divisions on the ocular micrometer with oil immersion objective?**

6. **How many divisions on ocular microscope would a single celled organism that was 240 µm take up oil immersion objective?**

7. If the magnification of an eyepiece is 10X and the objective is 40X, explain the magnification power of the microscope with three objectives?

8. Why do the microorganisms exist in different morphological forms?

PRACTICE 8

Wet-mount and hanging-drop techniques

8.1 Theoretical information

The living microorganisms can be magnified through microscope in the determination of size, shape, motility, and the other characteristics. Motile bacteria move with the help of flagella. A few bacteria have axial filaments through periplasmic space and these bacteria move with the help of these filaments but this movement cannot be detected in the microscope. Living bacteria cannot be colored, and very small. Nonstained preparation does not provide a sufficient degree of clarity, so that staining provides the clear visualization. Therefore, magnification of living microorganisms is difficult to magnify. Nonmotile bacteria have some vibrational movement in one direction due to the bombardment of flowing water molecules on bacterial cells. Kinetic energy of water movement causes movement of bacteria in one direction (Brownian movement). But this is not called microbial movement. Bacteria move quickly in all directions with flagella, but at the same time perhaps not all bacteria will move. The presence or absence of mobility of cells can be determined with the wet-mount and hanging-drop techniques. They can be determined the cellular shape and arrangement of cells (such as irregular clusters, packets, pairs, or long chains). Since the morphology of some preparations (spiral-shaped) is largely distorted by the staining method, the stain-free preparations allow true bacterial shape (spiral) to appear. Cellular changes during cell division can be determined. Germination of cells can be seen. Vacuoles, lipid granules, and cell inclusion bodies can also be detected with these techniques. Microscopic examination of microorganisms can also allow magnification of morphological appearance of cell: the size of cell and the natural shape of cell. During magnification, drying occurs gradually in the culture drop on the glass slide or hanging. The edge of the coverslip can be covered with a Vaseline ring to prevent drying. If the liquid culture is used, then place three or more loopfuls of the liquid culture (or other sample) directly on the glass slide. If the use solid culture is used, then first place one loopful of water onto the slide, place one or two loopfuls of culture on the glass slide and homogenize culture (colony) in the

Laboratory Practices in Microbiology
https://doi.org/10.1016/B978-0-323-91017-0.00016-0

water. Microbial cells from solid culture tend to cling to each other and cells must be dispersed sufficiently in the water; unless this is done, the smear will be too thick. Therefore, a very small amount of colony from solid culture is removed to make a good smear.

8.2 Microscopic techniques

Materials

Microbial cultures. 24 h tryptone soy broth cultures of *Escherichia coli*, *Klebsiella pneumonia*, and *Staphylococcus aureus*.

Waterandequipments. Steriledistilledwater,microscope,hallow-ground depression slide, microscope glass slides in bottle containing alcohol, cover slips, forceps, loop, and Vaseline.

With each culture provided, perform the steps has been described in the following techniques.

8.2.1 Hanging-drop technique

Hanging drop preparation is given in Fig. 8.1. Perform the following procedure for each bacterium.

1. Remove cover slip with forceps, deep into alcohol, and slightly flame to disinfect.
2. Write the initials of the microorganism and your name on the left-hand side of the depression slide with a waterproof marking pen.

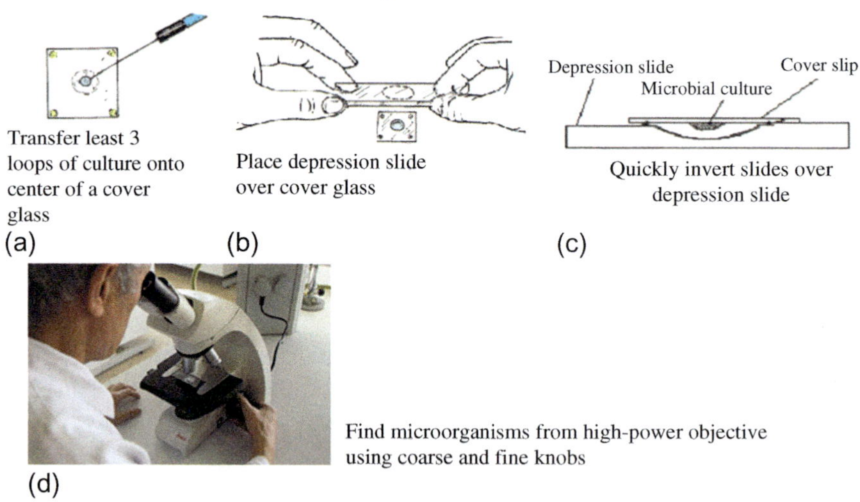

Fig. 8.1 Hanging drop preparation and magnification of preparation.

3. Place a small amount ofVaseline (or a loopful of water) to the ends of the cover slip (Fig. 8.1a).
4. Flame the loop to redness. Remove the plug of culture tube and flame the lip of the tube.
5. Place three or four loopfuls of a microbial suspension on the canter of cover slip (Fig. 8.1a).
6. Flame the lip of the tube and replace the plug. Also flame the inoculating loop before putting it down to receptacle.
7. Take a depression slide with forceps, immerse into alcohol in the bottle, flame the slide through Bunsen burner flame to burn alcohol to disinfect and allow the slide to cool for 30 s.
8. Label slide with the name of the microorganism and your initials using waterproof marking pen.
9. Place the depression slide over cover slip by placing concave area onto culture, slightly apply press on to the slide, and sure the cover slip and depression slide contact to each other (Fig. 8.1b).
10. Turn both slides in right side up readily on the depression slide and see the droplets of bacterial culture on the hole of the depression slide as hanging (Fig. 8.1c).
11. Place the slide on the microscope stage (Fig. 8.1d), then do as follows:
    - Examine the culture first with the low-power objective using coarse and fine knobs.
    - Once the microbial cell is magnified using low-power objective, swing the high-power objective into position and readjust the appearance using fine knob (Fig. 8.1d). Focus near the drop edge, since most bacteria are drawn to the drop edge by surface tension.
    - Avoid delay in examination, since condensation of water may decrease clarity and microbial cells become less motile.
12. Determine type of motility and shape of microbial cells. Draw microbial cells in your laboratory notebook. Do as follows:
    - *Real mobility of microbial cell.* Microbial cells that can perform motion in different directions in the same area several times should be regarded as motile.
    - *Brownian movement.* The Brownian movement is the movement of microbial cells that occurs as a result of vibration caused by invisible molecules, such as water, which bombards bacterial cells. This movement is vibrational and unidirectional, in which case the microorganism is recorded as not-motile.

- If you see only a few cells showing motility, then the microorganisms must be determined to be motile. Characteristically, only a few motile cells can move at a given moment.

8.2.2 Wet-mount technique

Wet-mount preparation is given in Fig. 8.2 for each microorganism. Perform following procedures for each microorganism:

1. Remove microscope slide from bottle containing alcohol and pass through flame.
2. Write the initials of the microorganism and your name on the left-hand side of the slide with a waterproof marking pen.
3. Flame the inoculating loop to redness. Take culture tube, remove the plug of culture tube, and flame the lip of the tube.
4. Remove least three loopfuls of broth culture and place on to the center of a microscope glass slide (Fig. 8.2a).
5. Flame the lip of the tube and replace the plug. Also flame the inoculating loop before putting it to receptacle.
6. One end of a cover slip is placed in contact with the culture (Fig. 8.2b) and cover slip is gently lowered until it contacts with the slide (Fig. 8.2c).
7. Place the slide onto the microscope stage (Fig. 8.1d).
8. First, magnify the slide using the low-power objective using coarse and fine knobs.
9. When microbial cells are visible under low-power objective, swing the high-power objective into magnification position and readjust the magnification using fine knob (Fig. 8.1d). Since most bacteria may present at the drop edge due to the surface tension, magnify near the end of the drop.
10. Avoid delay in examination, since condensation of water may decrease clarity and microbial cells become less motile.

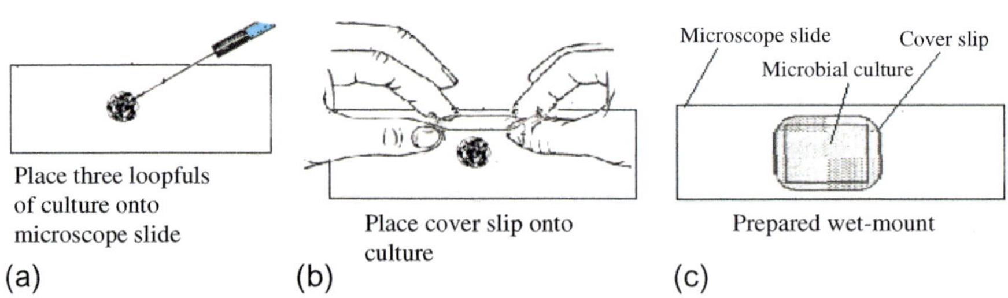

Fig. 8.2 Wet-mount preparation.

11. Determine type of motility and shape of microbial cells. Draw microbial cells in your laboratory notebook. Do as follows:
    - *Real motility of microbial cell.* Microbial cells that can perform motion in different directions in the same area several times should be regarded as motile.
    - *Brownian movement.* The Brownian movement is the movement of microbial cells that occurs as a result of vibration caused by invisible molecules, such as water, which bombards bacterial cells. This movement is vibrational and unidirectional, in which case the microorganism is determined not-motile.
    - If you see only a few cells showing motility, the microorganisms must be determined to be motile. Characteristically, only a few motile cells can move at a given moment.

8.2.3 Preparation from solid culture

Hanging-drop and wet-mount can also be prepared from solid culture. But better one is liquid culture. Since some of the cell structure like flagella will be distorted during preparation from solid culture. First, place a loopful of water on the slide and then aseptically transfer a very small amount of the solid culture from agar culture onto the slide near the water droplet with the loop. Spread the culture on slide by mixing with water to the size of the labeled circle. Reflame the loop after it has touched the slide. Follow the remaining procedure, follow as indicated in Section 7.2.1 or Section 7.2.2.

LABORATORY REPORT

Date:
Name:

Lab Title: Wet-Mount and Hanging-Drop Techniques

1. **Indicate name of bacteria, draw the bacteria you observed and record morphology: shape, motility, and0 purity of culture for:**

Wet-mount preparation:

Hanging-drop preparation:

2. **Which one is better in the morphological analysis of microbial cells: wet mount or hanging drop? Why?**

3. **Discuss aids of direct microscopy in the morphological analysis of microbial cells:**

4. How does true motility differ from Brownian movement?

5. Why are microorganisms hard to see in wet preparations?

6. Why the oil immersion lens is not used in examination of hanging drop procedure?

PRACTICE 9

Smear preparation and simple staining techniques

9.1 Theoretical information

Morphological characteristics of microorganisms can be observed with either the nonstaining (living) technique or staining (death) technique. Successive microbial staining depends first on the preparation of a suitable "smear" from the microbial culture. A good smear should be neither too thick nor too thin. If a smear is too thick, this means that too many cells have been put on the slide. A thick smear does not allow the proper penetration of light through the smear, and the morphological analysis of the individual cells cannot be determined. On the other hand, microbial cells are extremely small so if only a few cells are deposited on the slide, they are very difficult to find. Thick smear is possible when cells are obtained with a loop from solid culture. Thin smears usually result when the smears are made from culture. Dirt, dust, and other debris on the slide can be mistaken for microbial cells. That is why, you must be sure that you are working with a very clean slide if you want to prepare a "good smear."

The first step of staining is the smear preparation from liquid or a solid culture. If the liquid culture is used, place three or more loopfuls of the liquid culture (or other sample) directly on the glass slide. If the use solid culture is used, first place one loopful of water onto the slide, place one or two loopfuls of culture on the glass slide and homogenize culture (colony) in the water. Microbial cells from solid culture tend to cling to each other and cells must be dispersed sufficiently in the water; unless this is done, the smear will be too thick. Therefore, a very small amount of colony from solid culture is removed to make a good smear.

The chemical compounds used to stain bacterial cells are called dyes. Stained microbial cells are more readily visible under the microscope than the nonstained cells since nonstained cells are practically transparent and collect more light. Therefore, in the magnification of microorganisms in the microscope, stained bacteria are most easily observed. The use of a single stain to color, a staining is commonly referred to as simple staining. The simple staining involves the use of methylene blue, safranine, crystal violet

Laboratory Practices in Microbiology
https://doi.org/10.1016/B978-0-323-91017-0.00017-2

or acid fuchsin. Most cells and structures will stain the same color with the simple staining. Simple stains reveal characteristics of size, shape, and cell arrangement. Microbial cells have color-bearing ions (chromophores) that are positively charged (cationic) or negatively charged (anion), therefore they easily stained with ionic dyes.

The dyes used in the microbiology laboratory to stain the microbial cells are mostly acidic or basic. Acidic dyes (such as acid fuchsin, eosin) stain the basic structures of the cell (such as cytoplasmic contents, proteins, etc.) with a strong affinity. Because these structures of cells contain positively charged compounds and they are stained with negatively charged dyes. Basic dyes (such as crystal violet, methylene blue) have a strong affinity for acidic structures of the cells. The surface of a bacterial cell (such as cell wall) often contains a carboxyl group (−COOH), they are acidic negatively charged compounds and are stained with basic positively charged stains.

Dyes are commercially prepared as salts. The staining of bacteria is an ion-exchange reaction. In this reaction, positive and negative charges exchange between molecules from dyes and cell structure to form an ionic bond. For example, methylene blue (MB) is present as MB chlorine (MBCl) complex. When dye is rehydrated, MBCl ionizes to have a positive charge on the colored part of the molecule in solution, that is, MB^+. MB^+ molecules ionically bind to negative charges at the structure of the bacterial cell, when this happens, the cell is stained. This staining process can be written as follow:

$$\text{Bacterial cell-Na} + \text{ClMB} \rightarrow \text{Bacterial cell-MB} + \text{NaCl}$$

9.2 Staining techniques

Materials

Microbial cultures. 48 h cultures of *Bacillus subtilis*, *Escherichia coli* and *Staphylococcus aureus* in nutrient broth/agar.

Stains. Simple stains (methylene blue, crystal violet and safranine),

Equipments. Microscope glass slides in bottle containing alcohol, loop, needle, microscope, forceps, staining rack, wash bottle, blotting (bibulous) paper, xylene, and immersion oil.

9.2.1 Smear preparation

With each culture provided, perform following smear preparation procedure and as given in Fig. 9.1.

1. Remove microscope glass slide from slide bottle containing alcohol with forceps. Pass through the Bunsen burner flame to burn off alcohol and disinfect. Allow the slide to cool.

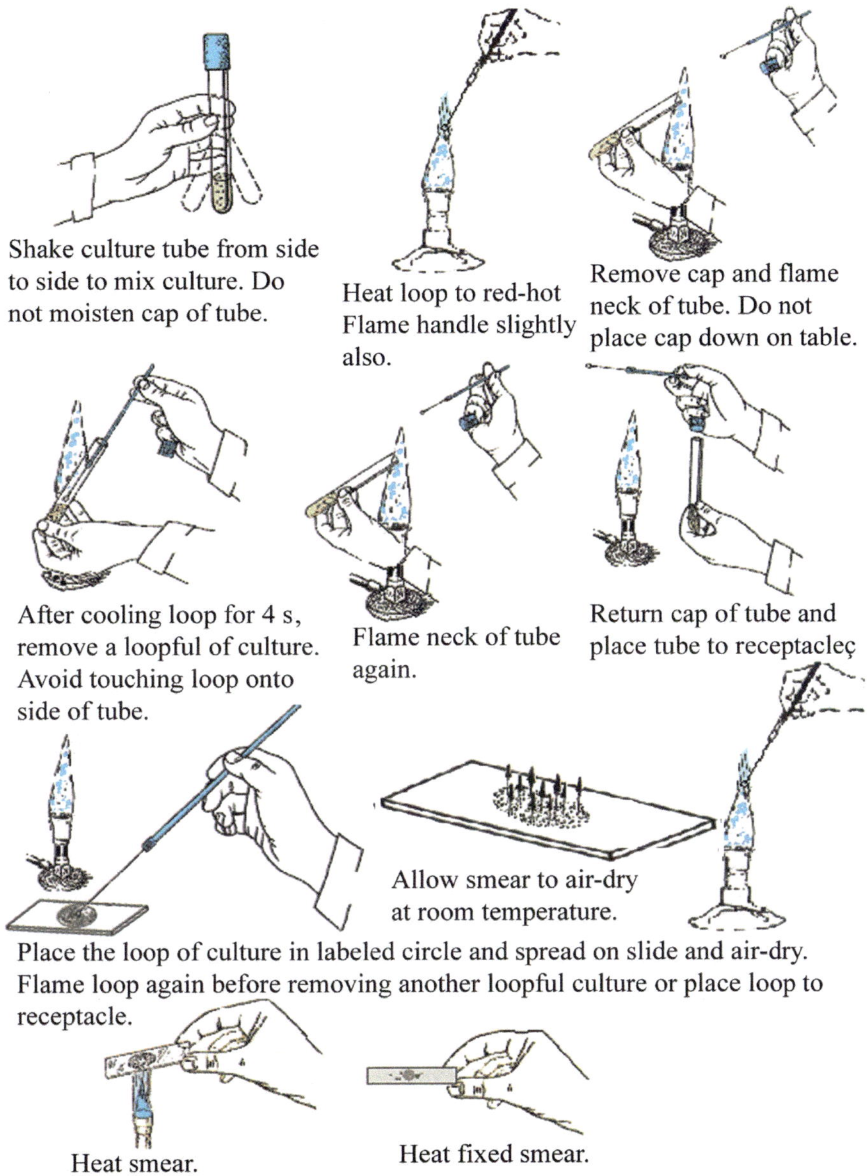

Fig. 9.1 Preparation of a microbial smear and heat fixation of smear.

2. Write the initials of the microorganism and your name on the left-hand side of the slide with a waterproof marking pen.
3. Make a 1.5 cm circle on the bottom of the slide, centrally located, with the marking pen.
4. Take the loop with other hand. Sterilize the loop at 90 degree angle through Bunsen burner flame and heat the entire wire portion to redness, flame handle slightly also and allow the loop to cool for 4 s near flame.

5. Take a tube of broth culture with left hand and carefully shake the culture vigorously.
6. Remove the cap of the tube with the little finger of the hand holding the loop. Flame the neck of the tube by passing it quickly through the upper cone of the blue flame.
7. Insert the sterilized loop and remove one loopful of culture. (Attention. You should make sure that you cool the loop completely before touching the microbial culture. When it touches the culture on a hot essence, it spreads the microorganism cells around and the bacteria move to the air and the surface. This situation also makes an irritating scratchy noise in the liquid environment This is also kills or mutate microbial cells).
8. Flame the mouth of the tube again and replace the cap. Put the tube to receptacle.
   (a) For liquid culture, place 3–5 loopfuls of broth culture on the slide and spread broth culture out to labeled circle. Carefully reflame the loop after it has touched the slide. Allow the broth culture to air dry. Don't apply heat.
   (b) For solid culture, first place a loopful of water on the slide and then aseptically transfer a very small amount of the solid culture from agar culture onto the slide near the water droplet with the loop. Spread the culture on slide by mixing with water to the size of the labeled circle. Reflame the loop after it has touched the slide.
9. Allow the slide to air dry. Don't apply heat.
10. After the film has become completely dry (called smear), rapidly pass the smear (film side up) through the blue portion of the Bunsen burner flame three times (heat fixation) to heat-kill microbial cells, fix them onto the slide, increase permeability of cell to dye and increase staining reaction of cell components. Slide should be just warm to your hand (Do not overheat!).
11. Cool the slide near flame, it will happen quickly within a few second. Do not pour stain on to hot slide. The heat-fixed smear is now ready for staining.

9.2.2 Simple staining

With each prepared smear, perform the simple staining steps described in this procedure and the procedure outlined in Fig. 9.2.

1. Flood the smear with dye (such as methylene blue) and allow it to remain for 2 min.

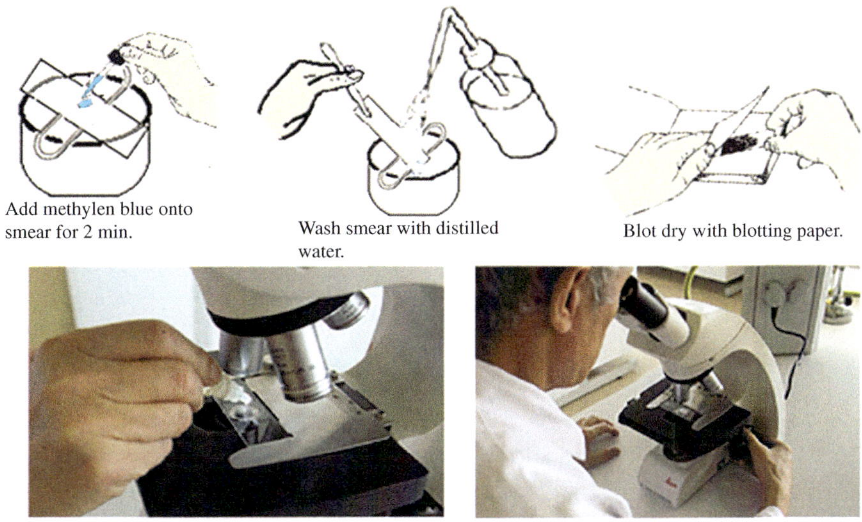

Add methylen blue onto smear for 2 min.

Wash smear with distilled water.

Blot dry with blotting paper.

After finding mocrobial cells from low power and high power objectives, simply rotate the high power objective out of position, swing oil-immersion objective one-half, add immersion oil onto stained area of slide and swing oil-immersion objective into position.

Fig. 9.2 Simple staining and magnification of stained smear.

2. Hold the slide at 45 degree with forceps. Pour the stain and rinse under gently running distilled water from wash bottle. Do not directly pour water on the preparation. Water should be poured from the top of the slide to allow it to flow over the slide.
3. Blot dry excess water with a piece of clean blotting paper. Do not rub. Air dry the slide.
4. Once the slide is placed on the microscope stage, do as follows:
   - **(a)** Examine the slide and find microbial cells first with the low-power objective using coarse and fine knobs.
   - **(b)** Once the image is visible under low-power objective, swing the high-power objective into position and readjust the appearance using fine knob.
   - **(c)** To use oil-immersion objective simply rotate the high-power objective out of position. Swing oil-immersion objective one half to working position. Add small amount of immersion oil (one drop) onto the stained area of slide and swing the oil-immersion objective into position and readjust appearance using the fine knob.
5. Draw several representative cells from each preparation, label each drawing with the name of microorganism and indicate the name of stain used. Indicate the arrangements, shape, color, and relative sizes of the cells (Fig. 9.3).

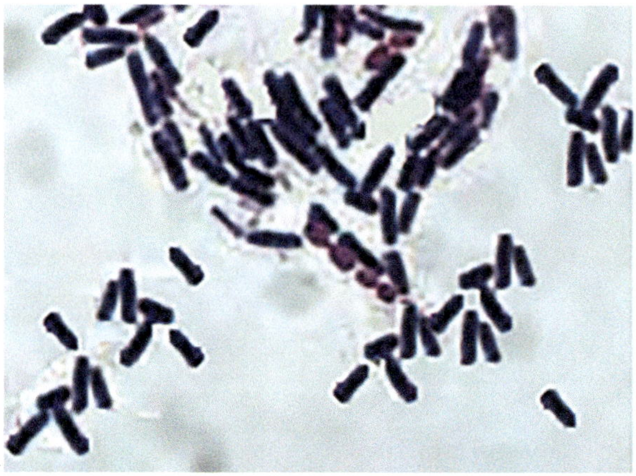

Fig. 9.3 Representative appearance of simple stained bacterial cells under microscope.

LABORATORY REPORT

Exercise Date:

Name:

Lab Title: Smear Preparation and Simple Staining Techniques

1. **Make drawings of stain preparations and indicate color, shape, and appearance of cell.**

			Unknown appearance
Stain of	Stain of	Stain of	Stain of

2. **Explain importance of smear preparation from solid and liquid cultures.**

3. **Explain importance of heat fixation.**

4. **Why is it necessary to use oil in conjunction with the 100X lens but not the other objectives?**

5. **What is the purpose of heat fixation?**

6. **How does smear preparation of cells from a liquid medium differ from preparation of cells from a solid medium?**

7. **Explain importance of staining of microorganisms.**

8. **What is the importance of living and death microorganism's analysis with light microscope.**

9. **What would happen if you apply too much heat while heat fixing the smear?**

10. **Why is it necessary to cool the loop prior to picking the culture?**

11. **Why is it customary to flame the inoculating needle, flask or tube mouth immediately prior to and after inoculation?**

12. **Do all bacteria stain with simple stains?**

PRACTICE 10

Gram staining technique

10.1 Theoretical information

Gram staining is a differential staining method since more than one dye is used as primary and secondary dyes. In the gram staining, the primary dye is crystal violet (CV), followed by the addition of an iodine (I) solution. Crystal violet and iodine forms an insoluble colored compound (CV-I complex) and iodine is called mordant. After decolorization, usually with 95% ethanol, a safranine counter stain is applied to the smear. The steps in the gram staining procedure and results of each step is given in Table 10.1. Crystal violet causes both gram-positive and gram-negative microorganisms turn to purple. When iodine solution is applied to cells, gram-positive and gram-negative bacteria remain purple. The function of iodine in gram staining is that it combines with crystal violet to form a relatively insoluble compound in the bacterial cells. When the decolorizing agent is added on purple cells, gram-negative bacteria lose their CV-I complex and become colorless after washing, but gram-positive bacteria retain the CV-I complex and remain purple. In the last step, safranin (secondary dye) is added to the bacterial cells, the color of the purple gram-positive bacteria isn't affected from safranin, and the discolored gram-negative bacteria are dyed pink with safranin.

There are factors affecting gram staining:

1. If the smear is overheated during fixation, the cell wall will burst, the gram-positive cells lose the CV-I complex during washing and appear pink with a secondary dye (safranin).
2. If crystal violet is added on a hot slide, it will be settled, the cells become impermeable to stain and cannot change color.
3. If the smear is insufficiently heat fixed, water wash out the stain from microbial cells during washing steps and cells become colorless.
4. Spreading a thick culture on the smear may not change color as fast during staining. Gram-negative bacteria be gram-positive.
5. A thick smear also causes the smear tend to "lift up" during staining.
6. Concentration and freshness of the gram staining reagents can cause variation in Gram reaction.

https://doi.org/10.1016/B978-0-323-91017-0.00028-7

Table 10.1 Steps in Gram staining and results of each step.

	Reaction and appearance of bacteria	
Solution application	**Gram-positive cells**	**Gram-negative cells**
1. Crystal violet (CV) (primary stain)	Cells stain violet color.	Cells stain violet color.
2. Iodine solution (I) (mordant)	Insoluble CV-I complex forming within cells; cells remain violet color.	Insoluble CV-I complex forming within cells; cells remain violet color.
3. Alcohol (decolorizer)	Cell walls dehydrate, shrinkage and small pores forming, CV-I complex can't pass out of cells during washing; cells remain violet color.	Cell walls dehydrate, shrinkage and large pores forming, CV-I complex passes out of the cells during washing; cells appear colorless.
4. Safranin (secondary stain)	Cells not affected and remain violet color.	Cells stained with safranine and become red color.

7. The amount of water remaining on the prepared smear can reduce reaction between crystal violet and iodine.
8. Too much washing after application of CV can reduce CV from cells, cells will be stained with safranin. Gram-positive bacteria appear as gram negative.
9. When old cultures of gram-positive bacteria are used in staining, they appear gram-negative with a tendency to change color during decolorization faster than normal. Approximately 16 h of cultures should be used in gram staining to obtain reliable results.
10. Some *Bacillus* species show gram variable; they are sometimes stained gram-positive and sometimes gram negative.
11. By flooding the slide with an excess amount of stain each time, you will prevent drying and consequent precipitation of dye on the slide, which can cause gram variables on the bacterial cells.
12. When the smear does not dry and is wet before heat fixation, heat causes the coagulation of proteins, which causes the gram-negative bacteria to appear gram-positive, since coagulation of protein will prevent decolorization of cells.

This technique will allow to place any bacterium into one of three broad groups. That is, the microorganism will be gram-positive, gram-negative, or gram nonreactive. The gram nonreactive includes some *Spirochetes* and acid-fast microorganisms (such as genus *Maycobacterium*). They are only weak gram positive and are better studied by other staining techniques.

10.2 Gram staining

Materials

Microbial cultures: 24 h nutrient broth cultures of *Escherichia coli*, *Staphylococcus aureus*, *Bacillus subtilis* and *Klebsiella pneumonia*.

Solutions: Crystal violet solution, safranin solution, iodine solution, alcohol (95%), immersion oil and xylene.

Equipments: Glass slides in bottle containing alcohol, pisette containing distilled water, blotting paper, loop, staining rock, forceps, and microscope.

10.2.1 Gram staining procedure

With each culture provided, perform following procedure and in Fig. 10.1:

1. Prepare a bacterial smear on slide for each microorganism as indicated in Section 9.2.2. Be sure to air dry and heat fix the smear, and label the slides carefully.
2. Cover the smear with crystal violet and allow 1 min.
3. Hold the slide at 45 degree angle with forceps, pour off excess stain, and wash off the stain from slide using distilled water the wash bottle. Do not directly add wash water onto smear, wash with flowing water from upper side of slide. Drain off excess water from slide.
4. Replace the slide on the staining rack, cover the smear with Gram's iodine solution and leave one minute.
5. Hold the slide at 45 degree angle with forceps, pour off excess iodine solution, and wash off the iodine from slide using distilled water from the wash bottle. Drain off excess water from slide.
6. Flood 95% alcohol for 10–20 s and only a faint color remains on the slide. When the dye flows from cells on the slide and cells become faint color, decolorization has occurred. Do not decolorize it too much.
7. Rinse immediately with distilled water from wash bottle for a few seconds. This will stop the decolorizing process.
8. Replace the slide on staining rock and cover the preparation on slide with safranin. Allow slide to remain for one minute.
9. Hold the slide at 45 degree angle with forceps, drain excess stain and immediately wash gently with distilled water from the wash bottle for few seconds.
10. Blot dry stained surface area of slide with a clean blotting paper to remove excess water and clean back of slide with blotting paper. Do not rub the surface area of slide. Dry the slide in air.

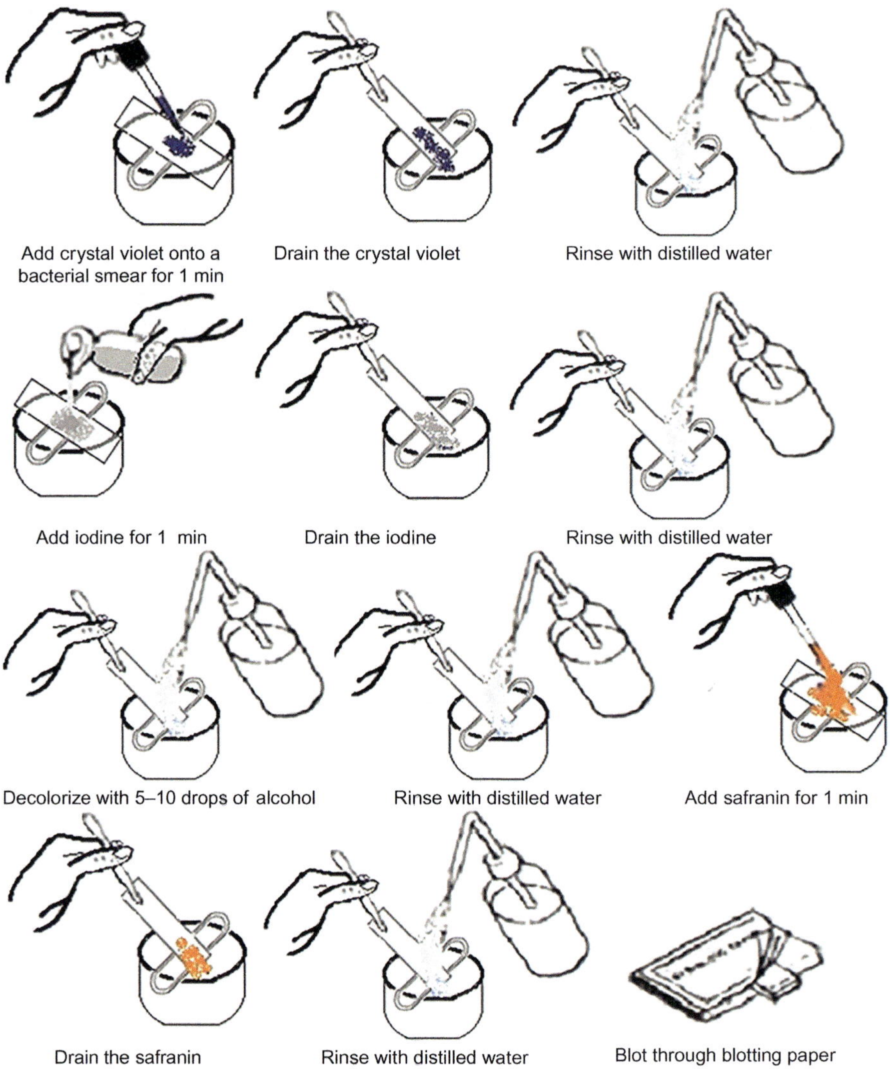

Fig. 10.1 Gram staining.

11. After the stained slide is placed on the stage of the light microscope, the following should be followed (Fig. 10.2):

(a) Examine the slide and find microbial cells first with the low-power objective with coarse and fine knobs.

(b) Once the image appears on the low-power, swing the high-power objective into magnification position and readjust the appearance using fine knob.

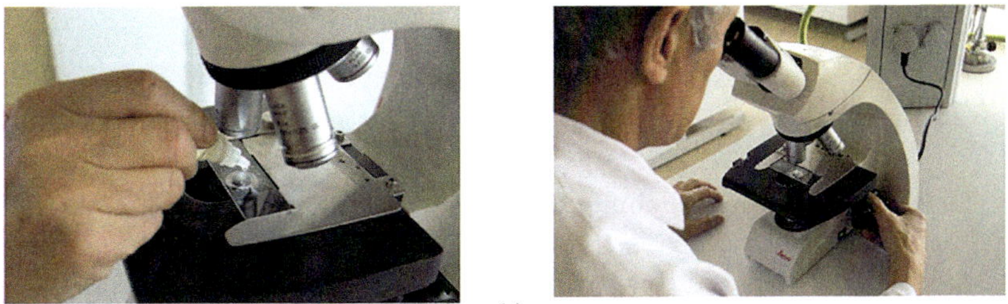

(a)

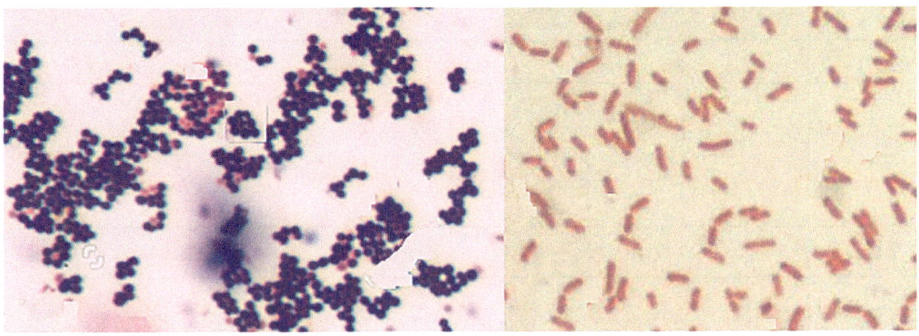

(b)

Fig. 10.2 Gram stained cells magnification (a) under microscope and (b) results of magnification for Gram-positive *(left)* and Gram-negative *(right)* bacteria.

   (c) To use oil-immersion objective simply rotate the high-power objective out of position. Swing oil-immersion objective one half to working position. Add small amount of immersion oil (one drop) onto the stained area of slide and swing the oil-immersion objective into position and readjust appearance using the fine knob.

12. Draw several representative cells from each preparation, label each drawing with the name of microorganism; indicate the gram reaction with color, the arrangements, shape, and relative sizes of cells.

LABORATORY REPORT

Date:

Name:

Lab Title: Gram Staining Technique

1. Make drawings of preparations and indicate characteristics of bacteria (such as color, shape, and appearance).

			Unknown
Stain of	Stain of	Stain of	Stain of

2. State why Gram staining is said to be a differential staining.

3. Describe the reasons of bacterial differentiation in the Gram staining procedure.

4. Explain the reasons why bacterial cells undergo gram.

5. Explain importance of Gram reactions.

6. **Explain how bacterial cells would look in the Gram staining procedure if the following mistakes were made:**
   (a) **Decolorizer left on too long**

   (b) **Decolorizer not left on long enough**

   (c) **Slide not heat-fixed before staining**

   (d) **What is the difference between a simple stain and a differential stain?**

7. **Explain why it is important to use only a small amount of culture when preparing a smear.**

PRACTICE 11

Bacterial endospore staining techniques

11.1 Theoretical information

Bacterial spores (such as *Bacillus*, *Clostridium*, and *Desulfotomaculum*) are small oval or spherical structures that are very resistant to high temperatures, radiation, desiccation, and chemical agents. When they are formed intracellularly, they are called endospore. The bacterial cell that gives the spore is called the vegetative cell. The spore is formed as a response to adverse conditions. Sporulation in bacteria is not a reproductive structure. Bacterial spore is a resistant structure to unfavorable environmental conditions. Each cell produces only one spore and each spore, in turn, germinate into one vegetative cell. The spore is just a part of the life cycle of spore forming bacteria under unfavorable conditions. The spores are metabolically inactive dormant form for years but are able to form vegetative cells (one cell per spore) in the suitable environmental conditions. Spores can be centrally, subterminally, or terminally located in the vegetative cell, and they can be larger or smaller in diameter than the vegetative cell. Large spore can cause swelling or enlargement of the vegetative cell. In some spore staining techniques, heat is used to ensure good penetration of the stain into the spore. Schaeffer-Fulton and Dorner techniques are used in the staining the bacterial spore. They can differentiate spores and vegetative cells. In Schaeffer-Fulton technique, primary dye malachite green is applied to a bacterial smear and smear is steamed over the flame to enhance penetration of the dye through impermeable spore coats. The Dorner spore staining technique can be used to dye bacterial endospores and vegetative cells. Heating is used in this staining technique and it increases penetration of stain. Nigrosine stain acts as a decolorizing agent as well as stain the background of a cell. After Nigrosine treatment it creates a clear contrast by removing the stain from cytoplasm. Endospores are stained during boiling, once the endospore is stained it doesn't get decolorize easily by water wash treatment. In this practice, you will learn how to make bacterial endospore staining techniques and how to view them under a microscope.

Laboratory Practices in Microbiology
https://doi.org/10.1016/B978-0-323-91017-0.00026-3

11.2 Endospore staining techniques

11.2.1 Schaeffer-Fulton spore staining techniques

Materials

Microbial cultures: 36 h trypticase soy broth culture of *Bacillus subtilis*.

Solutions and equipments: Schaeffer-Fulton staining set (malachite green and safranin), an electric hot plate, forceps, glass slides in bottle containing alcohol, loop, and blotting paper.

Procedure

With each culture provided, perform following procedure and the steps outlined in Fig. 11.1.

1. Prepare a smear of the microorganism and heat-fix a smear as follows:
   a. Place 1/2 drop of distilled water on a clean glass slide by touching the dropper to the slide.
   b. Sterilize the loop at 90 degree angle through Bunsen burner flame and heat the entire wire portion to redness, flame handle slightly also and allow the loop to cool for 4 s.
   c. Aseptically remove a small amount of the culture with loop from the bacterial colonies on the surface area of agar medium, place culture near water and mix the culture in the drop of water until it is cloudy. If too much culture is added to the water, you will not see stained individual bacteria and their spores.

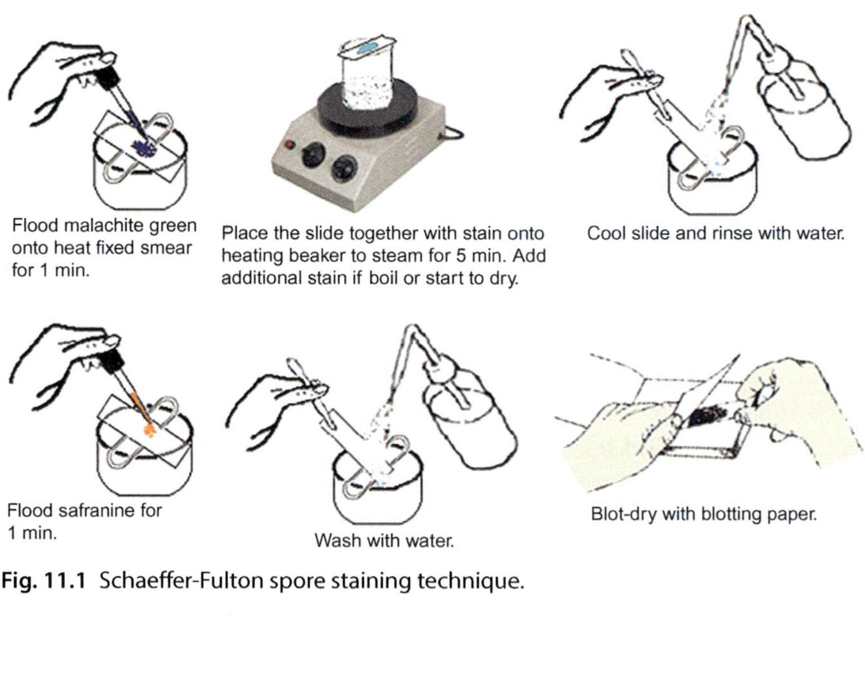

Fig. 11.1 Schaeffer-Fulton spore staining technique.

   - **d.** Spread suspension over the entire slide to form a thin film and flame loop.
   - **e.** Allow thin film to completely air dry.
   - **f.** Heat-fix the smear by passing the slide (film side up) through the Bunsen burner flame as many times as it takes until the slide just becomes uncomfortable to hold through fingers. Allow the slide to cool for a few seconds.
2. The slide is placed on the staining rack and flood malachite green stain onto the smear.
3. Allow the stain to react at room temperature for one minute.
4. Transfer slide onto the small beaker retaining boiling water (about 15 mL water).
5. Steam the slide over boiling water for 5 min. Do not boil or allow the stain to prevent dry. Add additional malachite green as it evaporates with steam from the slide (or hold the slide with forceps, the slide is carefully heated in the flame of a Bunsen burner until the stain begins to evaporate. When evaporation begins, the slide is pulled from the heat and left at the edge of the flame until evaporation stops; then it is slowly resteam).
6. Allow the slide to cool after heating, during cooling continue to add the malachite green onto the slide since stain continues to evaporate from hot slide.
7. After cooling the slide, pour excess stain from slide and wash the excess stain with slowly running distilled water for 30 s and blot the slide to dry.
8. Replace the slide on the staining rock and flood it with safranin for one minute.
9. Holding the slide at a 45 degree angle with slide forceps, pour off excess stain and wash off the excess safranin with distilled water.
10. Blot-dry slide with blotting paper carefully, clean back of slide with blotting paper and allow it to air dry.
11. Once the slide is placed on the microscope stage, do as follows (Fig. 11.2):
    - **a.** To find microbial cells under the microscope, first look at the preparation with a low-power objective using coarse and fine knobs.
    - **b.** When the microbial cells appear on the low-power objective, the high-power objective is changed with the low-power objective and the appearance of the microbial cells is readjusted using fine knob.

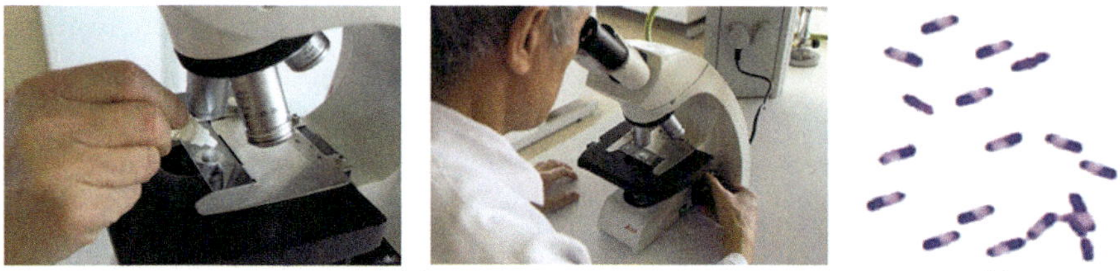

(a) (b)

Fig. 11.2 Magnification of stained culture (a) and appearance of endospores (b).

   c. Rotate the oil-immersion objective half way to the magnification position, add a drop of immersion oil (small amount) on the stained culture, and rotate the oil-immersion objective to the magnification position, and the appearance of the microbial cells is readjusted using fine knob.

12. Observe green spores and red vegetative cells under the microscope. Draw several representative cells and spores, label the spores and vegetative cells, and indicate their color.

11.2.2 Dorner spore staining technique

Materials

Microbial cultures. 24–36 h nutrient agar slant culture of *Bacillus cereus*.

Stains. Carbolfuchsin and nigrosine,

Equipments. Electric hot plate, dropper, small beaker (25 mL size), loop, small test tube (10–75 mm size), test-tube holder and glass slide in bottle containing alcohol.

With each culture provided, perform following procedure and the steps outlined in Fig. 11.3:

1. Sterilize the loop at 90 degree angle through Bunsen burner flame and heat the entire wire portion to redness, flame handle slightly also and allow the loop to cool for 4 s.
2. Prepare a suspension of bacterial culture by dispersing several loopfuls of culture from slants in 5 drops of sterile water in the small test tube.
3. Add 5 drops of carbolfuchsin onto bacterial suspension in the tube.
4. Place tube into the small beaker retaining boiling water (about 15 mL water).
5. Heat the suspension in the beaker of boiling water for 10 min.
6. Transfer 0.5 mL of cooled suspension onto one end of slide using pipette (1 mL) and add a drop of nigrosine onto the suspension.

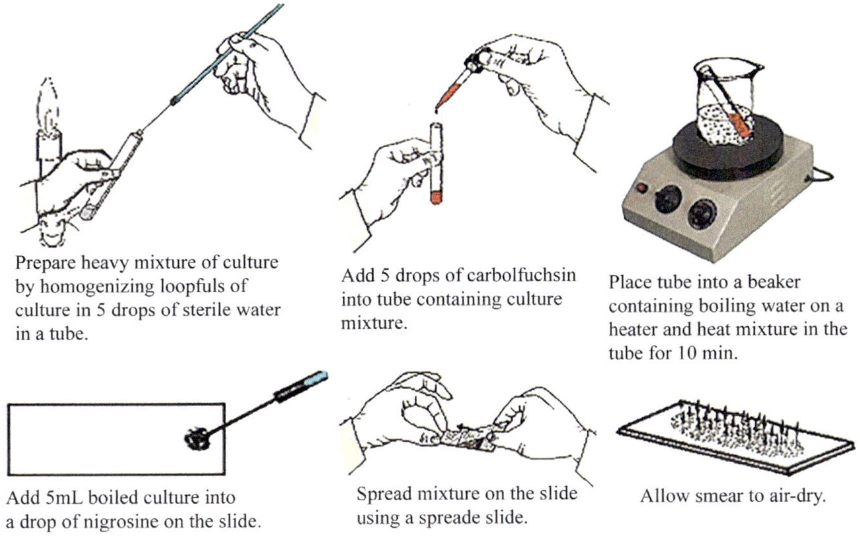

Fig. 11.3 Dorner spore staining technique. *(Spread mixture on the slide added with permission from McGraw-Hill; H.J. Benson, Microbiological applications: laboratory manual in general microbiology, 2001, Fig. 16.2, p. 68.)*

7. Spread the mixture on the slide using another slide by spreading from one end to other as in Fig. 11.3.
8. Allow the smear to air-dry.
9. Once the slide is placed on the microscope stage, do as follows (Fig. 11.2).
   a. To find microbial cells under the microscope, first look at the preparation with a low-power objective using coarse and fine knobs.
   b. When the microbial cells appear on the low-power objective, the high-power objective is changed with the low-power objective and the appearance of the microbial cells is readjusted using fine knob.
   c. Rotate the oil-immersion objective half way to the magnification position, add a drop of immersion oil (small amount) on the stained culture, and rotate the oil-immersion objective to the magnification position, and the appearance of the microbial cells is readjusted using fine knob.
10. Observe red spores and colorless vegetative cells. Draw several representative cells with spore, indicate shape of spore, label the spores, vegetative cells and their color. With this endospore staining procedure, endospores will stain red while vegetative bacteria will stain colorless.

LABORATORY REPORT

Date:
Name:

Lab Title: Bacterial Endospore Staining Techniques

1. **Make drawings from stained preparations and indicate characteristics (color, shape and appearance).**

Unknown

Stain of **Stain of** **Stain of** **Stain of**

2. **Name endospore-producing genera of bacteria.**

3. **State the function of bacterial endospores.**

4. **Explain importance of spore forming bacteria in food industry.**

5. **Do you think you would find more endospores in a freshly prepared culture or in an older culture of *Bacillus subtilis*? Explain.**

6. **Why do you think an infection caused by an endospore former might be harder to treat than one caused by a non-spore former?**

PRACTICE 12

Bacterial capsule staining techniques

12.1 Theoretical information

Most bacterial cells secrete a viscous substance that accumulates around the cell and coats the cell wall. This structure, depending on the thickness of the layer and its viscosity, is called capsule, microcapsule, or slime layer. The capsule serves a protective function for the bacterium by acting as a "barrier" between cell and environment. The virulence and infectivity of microorganisms are increased by the presence of capsule. *Klebsiella pneumoniae*, *Bacillus anthracis*, *Diplococcus pneumonia*, *Streptococcus pneumoniae*, *Haemophilus influenza*, and *Clostridium perfringens* are capsule forming bacteria. The bacterial capsule is usually composed of polysaccharides, polypeptides, or polysaccharide-protein complexes that are water soluble.

Capsule: Capsule is a discrete structure of polysaccharides deposited outside the cell wall. A capsular layer of extracellular polysaccharide material encloses bacteria and serves many functions. The presence of a capsule in *Streptococcus pneumoniae* is the most important factor in its ability to cause pneumonia. *Streptococcus mutans*, which causes dental caries, splits the sucrose in food and uses one of the sugars to build its capsule, which sticks tightly to the tooth.

Microcapsule: Microcapsule is polysaccharide structure mixed around the cells. It can be visible under electron microscope and has immunological significance. It is much thinner than the true capsule and cannot be demonstrated by light microscope is called microcapsule.

Slime layer: A slime layer is loosely associated structure on the cell wall of the bacterium and can be easily washed off, whereas a capsule is attached tightly to the bacterium and has definite boundaries. A glycocalyx is considered a slime layer when the glycol protein molecules are loosely associated with the cell wall. Bacteria that are covered with this loose shield are protected from dehydration and loss of nutrients.

Biofilm: It is a living ecosystem made of millions of bacterial cells, their wastes, and other extracellular products. This microbial biofilm can be found everywhere. This biofilm can easily reach a thickness with hundreds of cells

https://doi.org/10.1016/B978-0-323-91017-0.00012-3

on the surfaces of the equipments, teeth, and over time, plaque build-up can become mineralized, eventually forming calculus (tartar). Biofilms also have other serious medical implications. Persistent biofilms containing pathogenic bacteria can be problematic when they accumulate on damaged tissues, internal medical devices (such as catheters and pacemakers) and equipments used in the food production.

The bacterial capsule is nonionic, it can't be stained in the usual manner. The capsule is stained with a negative staining technique. A small amount of culture is mixed with India ink, and the resulting smear is stained with a dye, Loeffler's methylene blue, which penetrates the bacterial cell. Upon magnification with microscope, the background is dark, the capsule appears as clear unstained rings (negative stain), with small blue cell in the center of the rings (positive stain).

Gin's capsule staining technique is also used in the staining capsule. Because of its acidic nature, nigrosine (or India ink) stains the background dark. Nigrosine is recommended to India ink because it gives a more even background and spreads more easily. On the other hand, crystal violet is used for number of reasons including: To act as a fixative, increase penetration power, stain the cells (being a basic dye) and decrease pH of smear.

12.2 Capsule staining techniques

12.2.1 Negative capsule staining technique

Materials

Microbial culture: 48 h tryptose phosphate agar (TPA) slant cultures of *Klebsiella pneumoniae.*

Stains and equipments: Fresh nigrosine (or India ink), glass slides in a bottle containing alcohol, cover slips, forceps, and loop.

Procedure

From culture provided, perform one of the procedures as indicated in the following procedures and Figs. 12.1 and 12.2.

1. Take the loop from receptacle by hand. Sterilize loop over the flame of a Bunsen burner until all of the wire becomes redness. The entire wire must be heated and cool the loop for at least 4 s near flame.
2. Prepare culture with one of the following techniques:
   (a) **Spreader slide technique** (Fig. 12.1)
      i. Remove a glass slide from bottle and flame to disinfect the slide and allow to cool.

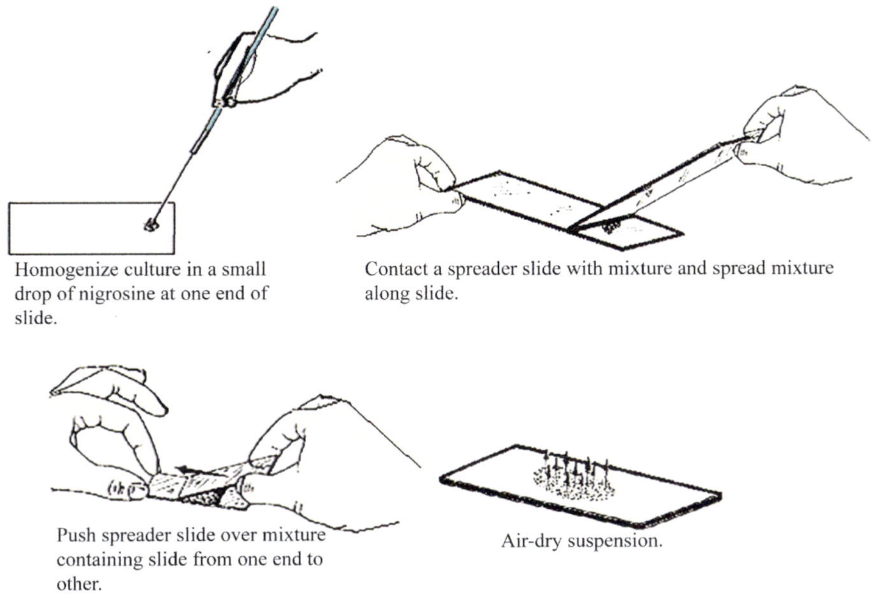

Fig. 12.1 Preparation of spreader slide in the negative staining technique. *(Modified with permission from McGraw-Hill; H.J. Benson, Microbiological applications: laboratory manual in general microbiology, 2001, Fig. 11.1, p. 56 and 11.2, p. 57.)*

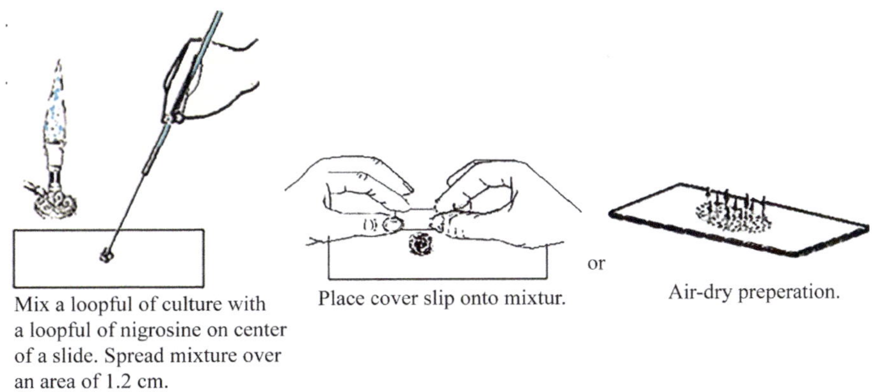

Fig. 12.2 Negative capsule wet-mount (or dry) preparation in negative staining technique.

ii. Using dropper, place one drop of nigrosine on one end of glass slide.

iii. Aseptically remove a loopful of culture, place near nigrosine and mix with the nigrosine (or India ink) within a 0.5 cm area in diameter on the slide. Flame and cool loop.

- **iv.** Remove a second glass slide from bottle and flame to disinfect the slide and allow to cool.
- **v.** Contact the second slide with suspension and spread suspension along spreading edge.
- **vi.** Push spreader slide over bottom to another end of slide.
- **vii.** Examine the preparation on the slide under microscope (or dry the suspension in air, not heat to dry and examine dried slide using microscope.

(b) Wet-mount technique (Fig. 12.2)

- **i.** Using dropper, place one drop of nigrosine (or India ink) on center of the clean slide.
- **ii.** Aseptically remove loopful of culture, place near nigrosine and mix with the nigrosine within 1–2 cm area in diameter on the slide until it is a uniform dark gray (not black). Flame and cool loop.
- **iii.** Carefully place a cover slip and examine wet mount under microscope (or dry the suspension in air, not heat to dry and examine dried slide using microscope.

3. After the slide is placed on the stage of the microscope (Fig. 12.3), the following is done: First, the slide is examined with a low-power objective. When the image appears under the low-power objective, the high-power objective is fulfilled and the magnification is adjusted using fine knob.

4. For dried preparation, to use oil-immersion objective, simply rotate the high-power objective out of position.

5. Rotate the oil-immersion objective half way to the magnification position, add a drop of immersion oil (small amount) on the preparation and rotate the oil-immersion objective to the magnification position, and the appearance of the microbial cells is readjusted using fine knob.

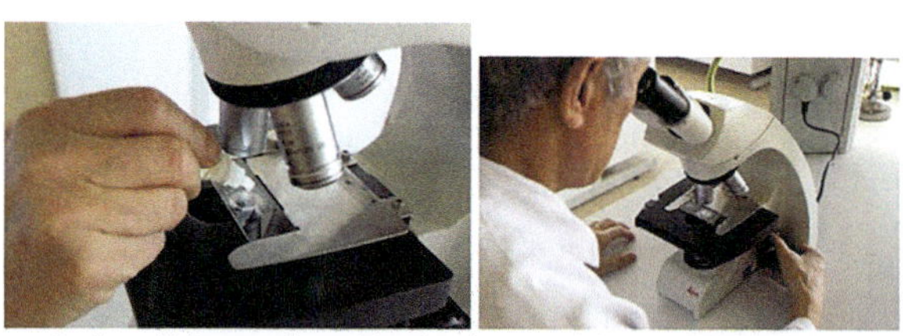

Swing oil-immersion objective into position and magnify.

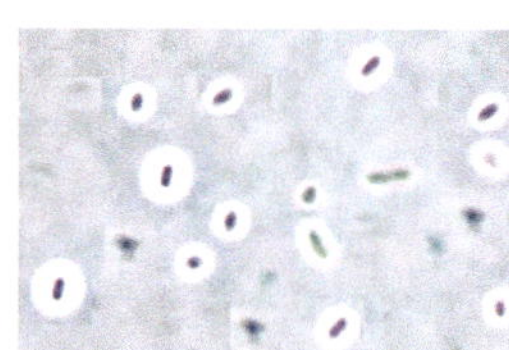

Capsule as a clear area (zone) around cell.

Fig. 12.3 Magnification of capsule from preparation under microscope and appearance of capsule around cells.

6. Observe capsule as a clear area (ring) around bacterial cells. Draw several representative cells, label the capsule, vegetative cells and their color.

12.2.2 Gin's capsule staining technique

Materials

Microbial cultures: 48 h tryptose phosphate agar (TPA) slant cultures of *Staphylococcus aureus* and *Klebsiella pneumoniae.*

Solutions and equipments: Fresh India ink (or nigrosine), Loeffler's methylene blue (or crystal violet), glass slides in a bottle containing alcohol, cover slips, forceps, loop and immersion oil.

Procedure

With culture provided, perform the steps described in this procedure and Fig. 12.4.

1. Remove the glass slide from the bottle containing alcohol with forceps. Pass through the Bunsen burner flame to burn off alcohol. Allow the slide to cool.
2. Sterilize the loop at 90 degree angle through Bunsen burner flame and heat the entire wire portion to redness, flame handle slightly also and allow the loop to cool for 4 s.
3. Transfer least four loopful of culture onto one end of slide.

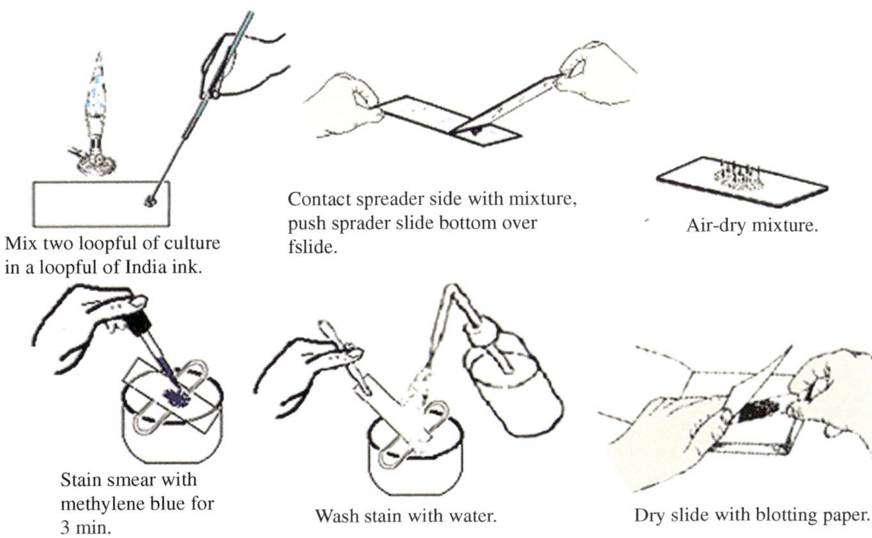

Fig. 12.4 Gin's capsule staining technique. *(Modified with permission from McGraw-Hill; H.J. Benson, Microbiological applications: laboratory manual in general microbiology, 2001, Fig. 14.1, p. 63.)*

4. Mix equal amounts of nigrosine and culture.
5. Take another microscope slide, lay one end on the first slide at an angle 45 degree touching the drop of mixture and use it to push mixture across the surface of the first slide. The smear will thus be covered with a thin film of dye.
6. Let the slide air dry for a few minutes. Do not blot the slide! Blotting will remove the bacteria from the slide and/or distort the capsule.
7. Flood the smear with Loeffler's methylene blue (or crystal violet), and let it for 3 min.
8. Wash with water and dry slide with blotting paper.
9. Examine the slide under microscope (Fig. 12.5). After the slide is placed on the stage of the microscope, the following is done: First, the slide is examined with a low-power objective using coarse and fine knobs. When the image appears under the low-power objective, the high-power objective is fulfilled and the magnification is adjusted using fine knob.
10. To use oil-immersion objective, simply rotate the high-power objective out of position.
11. Rotate the oil-immersion objective half way to the magnification position, add a drop of immersion oil (small amount) on the stained culture,

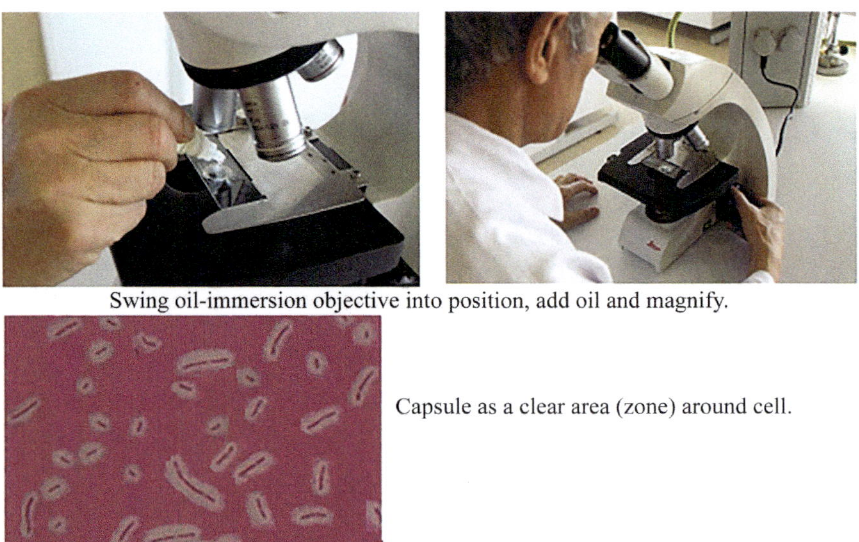

Fig. 12.5 Magnification of Gin's capsule stained preparation and appearance of capsules around cells.

and rotate the oil-immersion objective to the magnification position, and the appearance of the microbial cells is readjusted using fine knob.

12. Observe capsule around cells. Draw several representative cells with capsule, label the capsule, vegetative cells and their color, and indicate shape of cell. When viewed under the microscope, the background will be dark as a result of nigrosine, bacteria cells will be purple having taken the crystal violet dye while the capsule will be clear against a dark background given that it takes no stain. You may need to decrease the amount of light in order to make the capsule easier to see.

12.2.3 Examination of oral microorganisms

A small amount of material is taken from your teeth with a sterile straight toothpick or inoculation loop, and sample is mixed with the nigrosine on the slide and sample is homogenized in stain. Make sure that the microbial clusters in the sample must be completely homogenized. Sterilize loop before and after use, it is turned into a red core by starting with light heating. One of the capsule staining methods is applied to the preparation.

Caution. If you use a toothpick, discard it into a beaker of disinfectant.

LABORATORY REPORT

Date:

Name:

Lab Title: Bacterial Capsule Staining Techniques

1. **Make drawings of your stain preparations and indicate characteristics (color, shape and appearance).**

			Unknown
Stain of	**Stain of**	**Stain of**	**Stain of**

2. **Explain the chemical nature and functions of bacterial capsules.**

3. **Explain reasons of capsule formation by bacteria.**

4. **Explain importance of capsule for food industry.**

5. **How does the capsule contribute to organism's virulence?**

PRACTICE 13

Motility tests

13.1 Theoretical information

Many bacteria are motile with the help of flagella. They are protein in composition and form long threads which arise from the cell wall. Bacteria exhibit true or vital movement because of the presence of flagella. Brownian movement, on the other hand, is not because of the presence of flagella but this may be because of bombardment of the bacteria by the molecules of the suspending fluid. Flagella are very small in diameter; they cannot be seen directly by use of the light microscope. In order to see by light microscope, they must be stained. To observe flagella, a young bacterial culture must be used. Microbial motility can be detected using different microbiological techniques: wet-mount, hanging drop, staining, and motility medium. First two techniques are given in "Practice 8." In this practice, wet-mount, flagella staining and motility medium are given.

Two techniques for staining flagella are used: A wet-mount procedure (Ryu method) and dried-smear preparation (Leifson staining technique). The wet-mount and its staining is highly successful when a stable stain and regular slides and cover slips are used. This technique is simple for routine use when the number and arrangement of flagella are critical in identifying species of motile bacteria. The wet-mount and its staining preparations are not permanent for long time because the stain precipitates as the wet-mount dries.

In the wet-mount motility test, drops of cells were prepared in sterile water by using either of two nonstirring techniques: (i) touching the colony margin with an applicator stick (or wire) and then touching the stick into a drop of water on a clean slide; (ii) touching a loopful of water to the colony margin, allowing cells to swim into this loopful of water, and then touching the loopful of motile cells to a drop of water on a slide. Ryu stain were applied to the edge of the cover slip. To delay observation or to preserve the preparation overnight, the slide is held in a moist chamber. With each culture provided, perform the steps described in this procedure.

Motility test medium (MTM) is a semi-solid agar that can be used to demonstrate motility by diffusion of cells through medium. The medium contains lower amount of agar (0.5%) as well as triphenyltetrazolium

https://doi.org/10.1016/B978-0-323-91017-0.00030-5

chloride (TTC). TTC is a soluble compound which is taken up by bacterial cells. Once the substance has been absorbed by the bacteria, it is reduced. Reduction releases acid formazan and a highly pigmented pink color which is an insoluble compound. *Limitations*: Motility tests often show a false-negative reaction. The microorganism may be weakly motile or the flagella may be damaged because of heating, shaking, or other effects. A hanging drop motility may be performed from an inoculated tryptone broth incubated for 2–4 h to confirm motility results. TTC may be inhibitory to some bacteria. Most motility of bacteria may be interpreted at 35°C and *Yersinia enterocolitica* shows the best motility at 25°C.

13.2 Motility indication techniques

13.2.1 Flagella staining technique

Materials

Microbial cultures: 24 h brain heard infusion agar (BHIA) slant and BHIA plate cultures of *Proteus vulgaris* and *Staphylococcus aureus*. These cultures must be reinoculated and incubation three times with the use of fresh medium.

Solutions: Leifson flagella stain (solution A, solution B, and solution C), dichromat cleaning solution, 95% alcohol, 1% aqueous solution of methylene blue, sterile distilled water in tube.

Equipments: Sterile test tube, glass slides in bottle containing alcohol, pipettes (1 mL) in box, Pasteur pipette, and microscope.

13.2.1.1 Wet-mount motility test and Staining

1. Immerse slides in dichromat cleaning solution for about 1 week, wash thoroughly in distilled water, rinse in 95% alcohol, and allow draining of alcohol in air and dry the slide without wiping. Handle slides by their edges to avoid recontamination with grease.
2. Pass each slide back and forth through a flame for a few seconds.
3. Allow the slides to cool to room temperature with the flamed sides up.
4. Add 2 mL of sterile distilled water into BHIA slant using sterile 5 mL sterile pipette and wash off colonies from surface area of slant by gentle agitation.
5. Drops of motile cells were prepared in sterile water by using either of two nonstirring techniques:
   (i) *From BHIA slant.* Remove washed culture using 1 mL sterile pipette and add one drop of culture onto glass slide (Fig. 13.1).

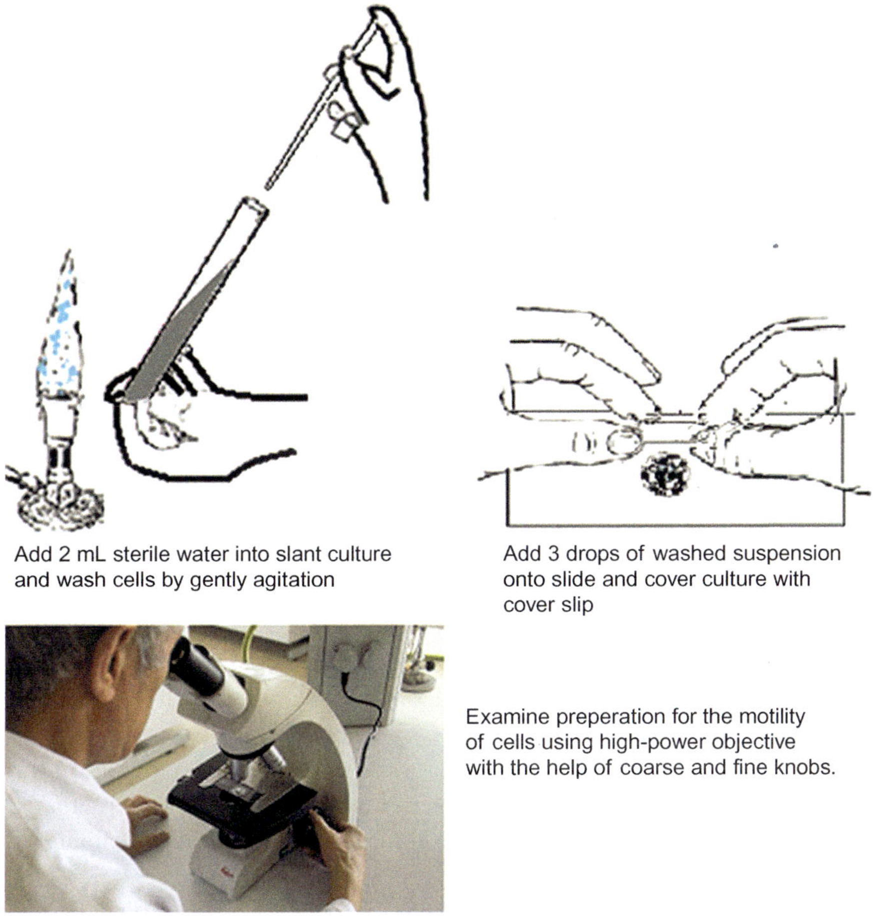

Fig. 13.1 Preparation of wet-mount and magnification under microscope.

(ii) *From BHIA plate***.** Touching the colony margin with the loop (or wire) on the BHIA plate and then touching the loop into a drop of water onto glass slide. Or touching a loopful of water to the colony margin on the nutrient agar plate, allowing cells to swim in the water and then touching the loop to the liquid near colonies. Place this loopful water–cell mixture onto glass slide.

6. Cover the faintly turbid drop of water on the slide with a cover slip. A proper wet-mount has enough liquid to fill the space under a cover slip. Small air spaces around the edge are preferable.

7. Examine the slide immediately under high-power objective (44X) for motile cells (Fig. 13.2). If motile cells are not seen, do not proceed with the stain.

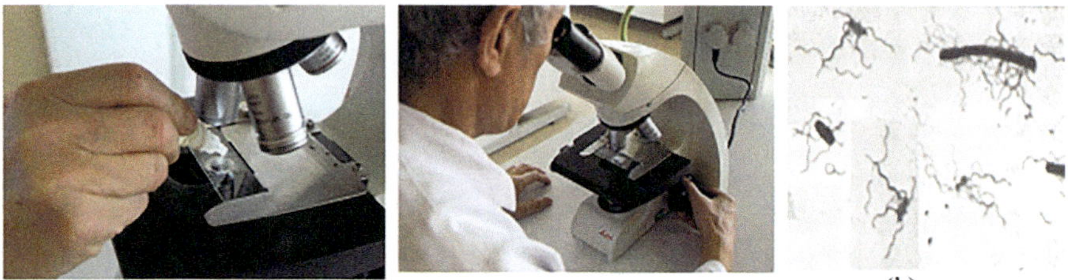

(a) (b)

Fig. 13.2 Magnification of flagella under microscope and appearance of flagella on bacterial cells.

8. If motile cells are seen, leave the slide at room temperature for 5–10 min. This allows time for the bacterial cells to adhere to either the glass slide or the cover slip.
9. Add two drops of Ryu stain gently to the edge of the cover slip. The stain will diffuse under the cover slip by capillarity action and mix with the cell suspension. Small air pockets around the edge of the wet mount are useful in aiding the capillary action.
10. The preparation is examined for flagella after 5–15 min allowing at room temperature. To delay observation or to preserve the preparation overnight, the slide was held in a moist chamber. Cells with flagella may be observed in the zone of optimum stain concentration, about half way from the edge of the coverslip to the center of the mount. Focusing the microscope objective (high-power) on the cells attached to the coverslip rather than the cells attached to the slide facilitates visualization of the flagella.
11. Observe the slide and note the followings: Presence or absence of flagella, number of flagella per cell and location of flagella on cell.

13.2.1.2 Leifson flagella staining

With each culture provided, perform the steps described in this procedure and Fig. 13.3.

1. Immerse slides in dichromat cleaning solution for about 1 week, wash thoroughly in water, rinse in 95% alcohol, and allow draining without wiping. Handle slides by their edges to avoid recontamination with grease.
2. Pass slide through a flame for a few seconds to burn alcohol and disinfect slide.
3. Allow the slides to cool at room temperature.
4. Use 22 h BHIA slant culture.

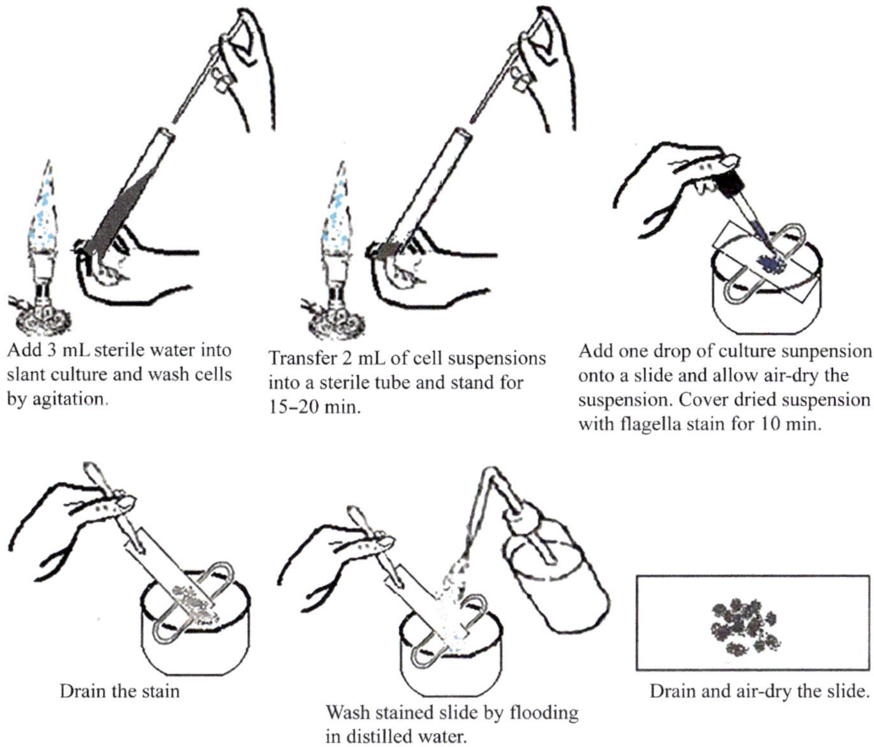

Fig. 13.3 Leifson flagella staining technique.

5. Wash off growth from slant by gentle agitation with 3 mL of sterile distilled water.
6. Transfer 2 mL suspension to a sterile test tube and allow the culture to stand undistributed for 15–20 min while flagella are regenerated and extended.
7. Remove a small drop (0.1 mL) with 1 mL pipette from top of the suspension (where motile bacterial cells are most numerous) and transfer to one end of the glass slide.
8. Tilt the slide to permit the liquid to flow down to the other end of the slide. If the drops fail to run, add another drop.
9. Allow to air-dry the suspension on the slide at room temperature. Do not heat fix!
10. Place the slide on a horizontal staining rack.
11. Mix leifson flagella stains A, B, and C in equal proportions. About 3 mL of each should be sufficient. The solution should be mixed fresh daily.

12. With the Pasteur pipette, quickly cover the smear with about 1 mL of the stain, and allow remaining for 10 min. The stain should not be allowed to flow outside the smear area. If stain dry, add stain onto the dried preparation on the slide.
13. Flood off the stain by adding distilled water to slide while it remains on the rack and drain the stain from slide.
14. Flood the smear with a 1% aqueous solution of methylene blue (or carbolfuchsin) down the slide for as short as time as possible (2–10 s).
15. Rinse counterstain by flooding with distilled water. Drain and air dry. Do not blot.
16. Place the slide under microscope (Fig. 13.2) and do as follows: Examine the slide first with the low-power objective using coarse and fine knobs. Once the image is visible under low-power objective, swing the high-power objective into position and readjust the appearance using fine knob.
17. To use oil-immersion objective, simply rotate the high-power objective out of position.
18. Rotate the oil-immersion objective half way to the magnification position, add small drop of immersion oil on the stained culture, and rotate the oil-immersion objective to the magnification position, and the appearance of the microbial cells is readjusted using fine knob. Observe flagella on the vegetative cells (Fig. 13.2).
19. Draw several representative cells; label the flagella, vegetative cells, and their color. The flagella stain red and the bacterial bodies, blue.

13.2.2 Semi-solid medium for motility test

Materials

Microbial cultures: 24 h nutrient agar slant cultures of *Proteus vulgaris* and *Staphylococcus aureus.*

Medium: Brain heard infusion broth (BHIB) Motility test medium (MTM, or sulfate indole motility medium, SIM).

Solution and equipment: 95% alcohol and needle.

Procedure

(a) Test tube technique

Inoculate tubes of MTM (or SIM) with each microorganism according to the following procedure:

1. Label the tubes of MTM (or SIM) medium with the name of the microorganisms and your initials.

2. Inoculate one loopful culture into 5 mL BHIB and incubate for 8 h.
3. Flame and cool the needle and insert it into the broth culture (4–8 h) after flaming the neck of the tube.
4. Remove the culture with needle and place the broth culture to receptacle after flaming the neck of the tube.
5. Remove the cap from the tube of MTM, flame the neck, and stab the needle 2/3 of the way down to the bottom of the medium, as shown in Fig. 13.4a. Flame the neck of the tube again before returning the cap to the tube. Place the tube to receptacle.
6. Repeat steps 2 and 3 for the other culture.
7. Incubate the tubes with the caps loose at 35°C for 24 h.
8. After incubation, observe growth of microorganisms in the media in the tube.
9. If bacterium is motile, they will diffuse into the soft medium laterally from the line of inoculation, resulting in a diffuse, pink color throughout the medium (Fig. 13.4c). Nonmotile microorganisms grow along the line of inoculation only, producing a pinkish-red color line with no diffusion (Fig. 13.4b).

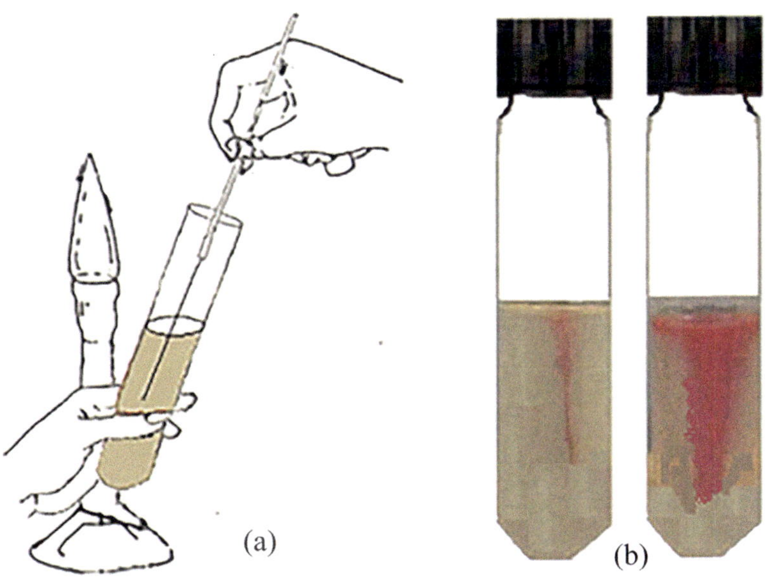

Fig. 13.4 (a) Inoculation into MTM by needle and (b) motility test results in MTM from *left* to *right*: motile and nonmotile.

(b) Plate technique

Mark the bottom of a plate of soft agar on the bottom of plate with two one cm circles about 2.5 cm apart. Label one circle for one of the microorganisms. Put your initials on the plate also. Remove Broth culture with needle. Inoculate by stabbing the needle into center of the circle on the MTM with *S. aureus* and the center of the other circle with *P. vulgaris.* Incubate the plate without inverting at 35°C for 24 h in the incubator. After inoculation, look at growth on the MTM on the plate.

Positive: A diffuse pink color occurring throughout the medium.

Negative: A pinkish red line at the circle site with no diffusion.

LABORATORY REPORT

Date:
Name:

Lab Title: Flagella Staining and Motility Medium Techniques

1. **Make drawings from your preparations and indicate characteristics (color, shape and appearance).**

			Unknown
Stain of	**Stain of**	**Stain of**	**Stain of**

2. **Draw the following flagellation arrangements: monotrichous, lophotrichous, amphitrichous and peritrichous. Indicate flagella arrangement from your result.**

3. **State the chemical nature and function of bacterial flagella.**

4. **How do you differentiate water current movement from true motility?**

5. **Explain importance of flagella and flagella staining.**

6. Draw the growth seen in each of two media after inoculation.

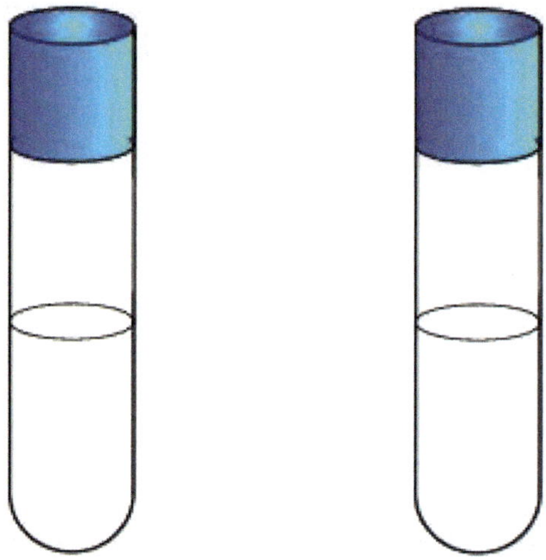

Semisolid medium culture of nonmotile bacteria

Semisolid medium culture of motile bacteria

7. Describe staining and motility medium of testing for the bacterial motility and indicate how to interpret the results.

SECTION III

Biochemical and cultural characteristics of microorganisms

Metabolism includes the degradative (catabolism) or the biosynthetic (anabolism) biochemical reactions. Both processes occur simultaneously and complement each other to provide for the essential needs of the cell. All metabolic reactions occur as a serial of steps through chemical compounds. Metabolic reactions are carried out through enzymes. In order to observe the microbial enzymatic reaction, it can be determined by cultivating a pure microbial culture in a special medium. There are differences between microorganisms in terms of enzymatic activities. These enzymatic reactions can be used to identify and differentiate microorganisms. Cultural characteristics of microorganisms depend on biochemical characteristics. Morphological, cultural, and biochemical characteristics are essential to identify the genus, species, or strains of microorganisms. Microorganisms are affected in different ways by their chemical and physical environments. In this section, some of the cultural, oxidation, fermentation, and hydrolytic reactions are explained through practices that will determine the biochemical and cultural characteristics of microorganisms.

The general practices given in this section are as follows:
Practice 14. Cultural Characteristics of Bacteria
Practice 15. Oxidation and Fermentation
Practice 16. Hydrolytic Reactions

PRACTICE 14

Cultural characteristics of bacteria

14.1 Theoretical information

The cultural characteristics of a microbial colony are analyzed on different kinds of media. The most used microbiological media in the cultural study are nutrient agar, nutrient broth, fluid-thioglycolate medium, and others. Certain risks arise when a culture is examined. Species variations in cultural characteristics can result. Variations can result from (1) genetically properties changes depending on environmental conditions, (2) spontaneous and undirected permanent genetic changes, known as mutations, and (3) formation of a new gene because of sexual reproduction. Five types of growth cultures can be used in the indication of cultural characteristics of bacteria in this practice: An isolated colony on the surface of an agar plate culture, nutrient agar slant culture, nutrient broth culture, fluid-thioglycolate culture, and gelatin deep culture. Different types of bacterial growth on these media are used in the indication of cultural characteristics.

14.2 Techniques used in the indication of cultural characteristics

Materials

Microbial culture: 24 h nutrient broth cultures of *Escherichia coli*, *Staphylococcus aureus*, and *Proteus vulgaris*.

Media: Nutrient agar plate, nutrient agar slant, nutrient broth, and nutrient gelatin deep.

Equipment: Loop, needle, ice-water bath, and incubator at 20°C and 35°C.

14.2.1 Inoculation of media

Perform following inoculations from each of the pure bacterial cultures.

1. Divide the bottom of the nutrient agar plate into two sections and label each sector with a given culture. Streak plate a loopful culture onto surface of its labeled sections.
2. Label all media in the tubes with microbial name, your initials, and type of medium.

Laboratory Practices in Microbiology
https://doi.org/10.1016/B978-0-323-91017-0.00014-7

3. Inoculate the nutrient agar slant surface starting from the bottom of the slant surface, moving the loop upward out of the slant area (not until upper end of agar medium) without streaking. Do not dig into the agar.
4. Inoculate nutrient broth with a loopful of culture.
5. Incubate all tubes and inverted plates at 35°C for 24 h in the incubator.
6. Make a stab inoculation with culture into nutrient gelatin stab by forcing a needle directly down into the medium to the bottom of the tube (until 1/4 cm of the bottom of the tube) and pulling it straight out, avoiding lateral motion. Tubes are incubated at 20°C for 24 h.

14.2.2 Evaluation of cultural characteristics

After the cultures are incubated under optimum growth conditions, they should be carefully moved to the table to avoid disturbing the growth pattern in the nutrient broth. The nutrient gelatin tube is placed in an ice water-bath. Investigation of microbial growth and activities in different media should be done as follows and record following proper descriptive terminology for the growth of cultures in your laboratory notebook.

(1) Nutrient agar slant culture

After incubation, examine the culture for growth on the slant to the following characteristics:

Amount of growth. The growth may be described as *none*, *slight*, *moderate*, and *abundant* according to the growth on the surface area of slant.

Color. Look for the pigmentation of the colony within the medium. Some microorganisms can produce soluble pigments that diffuse into the media. To be able to study diffused pigmentation, it must be tilted towards the light.

Opacity. Microorganisms on the surface of the slant will appear opaquer than the presence of a small growth on slant. Opacity degrees on the medium can be defined *opaque*, *transparent*, and *translucent* (partially transparent).

Form. The appearance of different types of colonial growth on slant is illustrated in Fig. 14.1. The differentiation of each type of microorganism can be defined by the following growth characteristics:

Filiform: characterized by uniform growth along the line of inoculation.
Echinulate: margins of growth exhibit toothed appearance.
Beaded: separate or semi-confluent colonies along the line of inoculation.
Effuse: growth is thin, unusually spreading.
Arborescent: branched, treelike growth.
Rhizoid: root-like appearance.

Fig. 14.1 Bacterial colony forms on nutrient agar slant from *left* to *right*: *filiform, echinulate, beaded, effuse, arborescent*, and *rhizoid*. *(Modified with permission from McGraw Hill; H.J. Benson, Microbiological applications: laboratory manual in general microbiology, 2001, Fig. 47.1, p. 157.)*

(2) Liquid culture

The terms shown in Fig. 14.2 are used to describe the various types of the growth characteristics of bacteria in liquid culture.

Surface. A *pellicle* type of surface differs on the surface area of broth. The *membranous* type of growth is much thinner. Bacterial cells floating as masses on medium can be expressed as a *flocculent* surface.

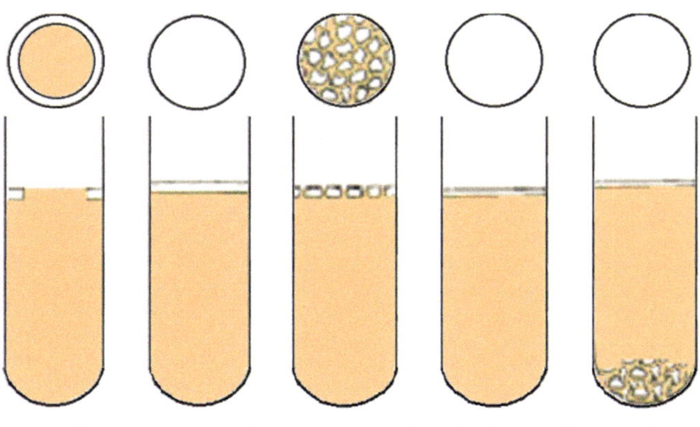

Fig. 14.2 Types of surface growth in nutrient broth from *left* to *right*: *ring, pellicle, flocculent, membranous* and *sediment*. *(Modified with permission from McGraw Hill; H.J. Benson, Microbiological applications: laboratory manual in general microbiology, 2001, Fig. 47.2, p. 158.)*

Subsurface. Below the surface of liquid medium, the cloudy appearance on the broth can be described as *turbid*, formation of small particles is called *granular* appearance; small masses with floating around the medium is known as *flocculent*, and large particles in suspension is known as *flaky*.

Sediment. The sediment amount in the tube bottom may vary from none to greater amount. The type of sediment can be described by agitating the tube for suspending the material. The sediment type can be described as *granular*, *flocculent*, and *viscid*. Test for viscosity of medium can be detected by inserting a sterile inoculating loop through the medium.

Amount of growth. The amount of growth by degree of turbidity is indicated by shaking the tube to disperse the microbial cells. Term, such as *slight* (scanty), *moderate* and *abundant*, describes the amount of growth in liquid medium.

(3) Gelatin stab culture

Nutrient gelatin stab culture is removed from the ice-water bath and examines it. Check first to see if liquefaction occurs. Microorganisms can liquefy gelatin production of enzyme gelatinase. If the growth occurs in nutrient gelatin, look for two things: The presence or absence of liquefaction. After the tubes have cooled for about 5 min in ice water, make your evaluation.

No liquefaction. Microorganisms will not grow in nutrient gelatin. Growth may occur without liquefaction. These results indicate non-gelatinase production (Fig. 14.3).

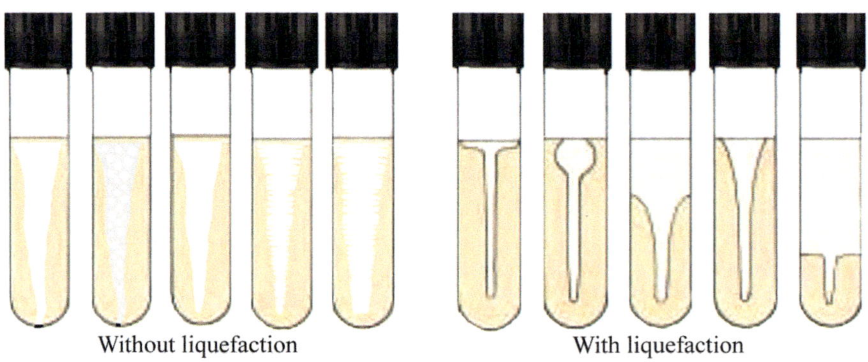

Fig. 14.3 Types of growth in gelatin stabs without liquefaction from *left* to *right*: *filiform, beaded, papillate, villous* and *arborescent* and with liquefaction from *left* to *right*: *crateriform, napiform, infundibuliform, saccate*, and *stratiform*. *(Modified with permission from McGraw Hill; H.J. Benson, Microbiological applications: laboratory manual in general microbiology, 2001, Fig. 47.3, p. 159.)*

Liquefaction. Some microorganisms produce the enzyme gelatinase, which can digest gelatin. Once the medium lacks gelatin, it loses its firm characteristics and liquefy the medium. Liquefaction may occur in different ways, as indicated in Fig. 14.3. The distinguishing characteristics of each are as follows: *Crateriform*: Saucer-shape liquefaction, *Napiform*: turnip like, *Infundibuliform*: funnel like or inverted cone, *Saccate*: elongated sac, tubular or cylindrical and *Stratiform*: liquefied to walls of tube in upper regions.

(4) Nutrient agar plate culture

Colony appearance on nutrient agar plates is described with respect to size, color, opacity, form, elevation, and margin. Indicate your results by using Figs. 14.4 and 14.5 for possible variations.

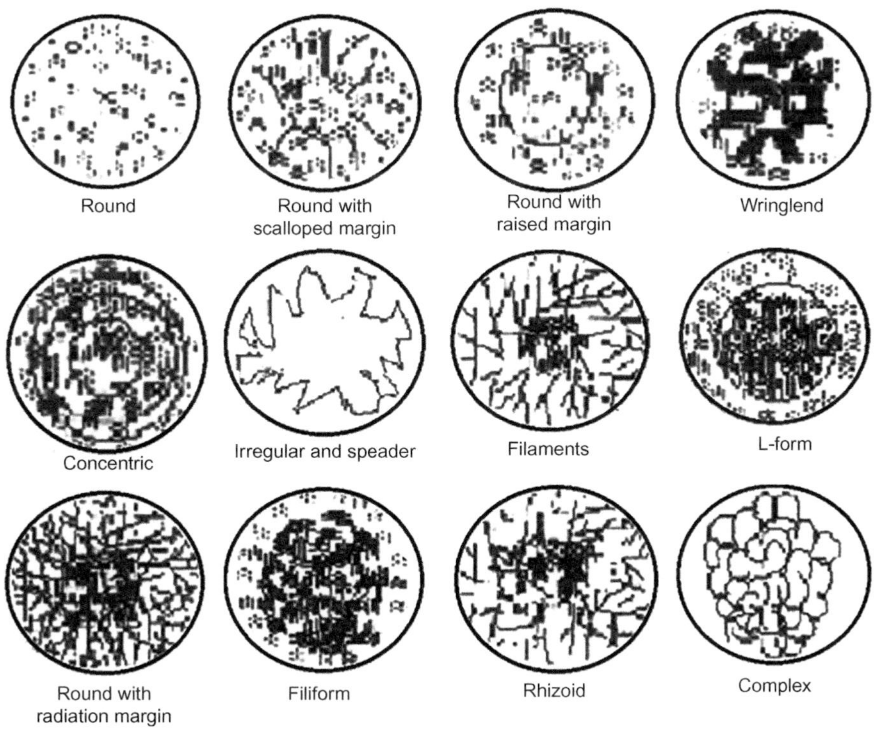

Fig. 14.4 Variation of colony appearance.

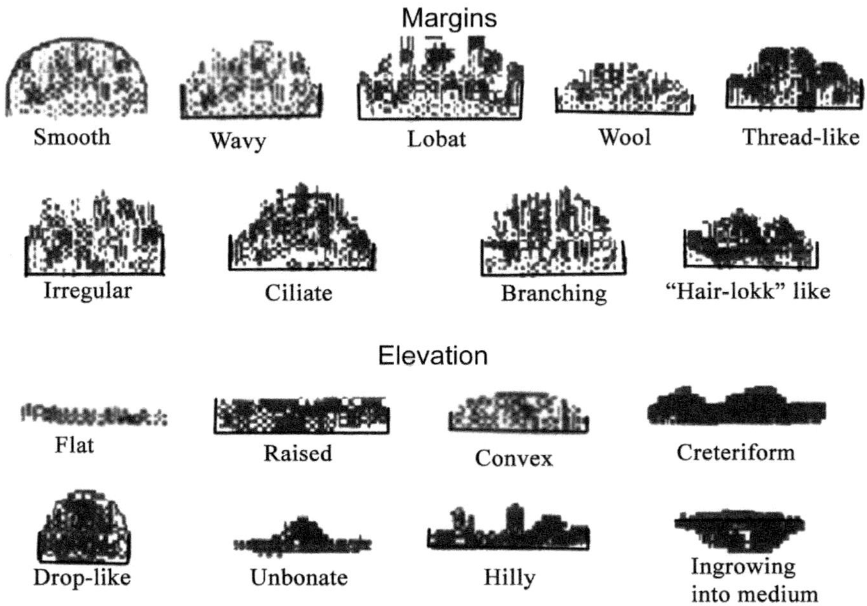

Fig. 14.5 Colony shape.

LABORATORY REPORT

Date:
Name:

Lab Title: Cultural Characteristics of Bacteria

1. **Record bacterial culture characteristics for all the cultures according to the growth in the following media.**

    (a) Nutrient agar slant:

    (b) Nutrient broth:

    (c) Gelatin stab:

    (d) Nutrient agar plate:

2. **Explain reasons of cultural characteristic variations among bacteria.**

3. **Discuss the importance of the study with cultural characteristics.**

PRACTICE 15

Oxidation and fermentation

15.1 Theoretical information

Microorganisms vary considerably in their ability to oxidize various carbohydrates. Some microorganisms can utilize one or more carbohydrates and produce acid and gas; others are able to produce acid but not gas; still others fail to ferment any carbohydrate. Such information is of considerable value in the identification and classification of microorganisms. All bacteria have endoenzymes (intracellular enzymes functional within the cell). Some bacteria also produce exoenzymes (extracellular enzymes functional in the outside of the cell). The exoenzymes are secreted out of the cell and diffuse into surrounding medium (Fig. 15.1). Carbohydrates are prime source of energy and carbon for synthesis of cellular substances. Simple sugars (such as glucose and galactose) are initiators of some metabolic reactions. Endoenzymes (such as maltase and lactase) decompose maltose and lactose (disaccharides), and yields glucose and galactose (monosaccharide), which are used in glycolysis. Some microorganisms utilize a carbohydrate slowly. Cellulose and starch are polymers of glucose. The starch molecule is first hydrolyzed to a smaller polysaccharide called dextrin. Amylase (exoenzyme) degrades the polysaccharide into disaccharides or monosaccharides that can be readily absorbed by the cell.

15.1.1 Metabolic reactions

The chemical reactions that occur in the living microorganisms by enzymes are called metabolism. Bacteria use oxidation or fermentation to produce energy in their metabolic systems. Oxidative microorganisms carry the electron among organic compounds in metabolic pathways and use oxygen as the final electron (as hydrogen) accepter and produce carbon dioxide (CO_2) and water as the final products. Fermentative microorganisms produce lower amount of energy, organic substances (such as acids, aldehydes, and alcohols), and gases (such as CO_2, hydrogen, and methane) from organic compounds. In fermentation, organic compounds act as both electron donor and acceptor.

Simple sugars, especially glucose, are compounds used primarily by fermentative microorganisms. Other organic compounds (such as acids, amino

Laboratory Practices in Microbiology
https://doi.org/10.1016/B978-0-323-91017-0.00001-9

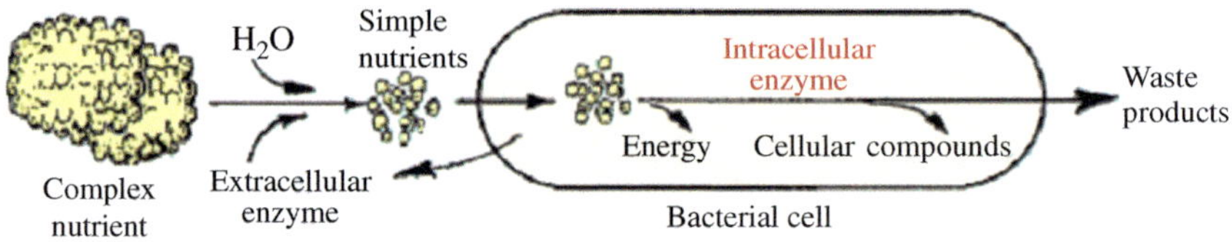

Fig. 15.1 Hydrolytic extracellular enzymes split larger substrate to simple nutrients. They are then absorbed by microbial cell and intracellular enzymes decompose smaller substrate to produce energy and cellular compounds.

acids, purines, and pyrimidines) can be fermented by some microorganisms. The final products of a fermentation vary depending on the nature of the microorganism, the properties of the substrate, and environmental conditions (such as temperature and pH). Some microorganisms can use both of fermentation and oxidation. An exception is observed in fermentation, where fermentation takes place in the presence of oxygen.

In the differentiation of enteric members (Table 15.1), biochemical tests can be used. Most used biochemical tests are IMViC (indicates: indole test, methyl red test, Voges-Proskauer test, and citrate test) and HOMoC (indicates: hydrogen sulfide production test, ornithine utilization test, motility test, and citrate test). Mixed acid fermenting bacteria yield lactic, acetic, succinic, and formic acids, CO_2, H_2, and ethanol. *Escherichia coli* decomposes formic acid to CO_2 and H_2 under acidic condition. In some enteric bacteria, sugar fermentation gives an additional major product, 2,3-butanediol, which is formed from pyruvic acid. This is the characteristic of *Enterobacter*, *Serratia*, *Klebsiella*, *Erwinia*, and certain members of Vibrionaceae (such as *Aeromonas*).

A significant number of gram-negative bacteria can be distinguished by their metabolic products as a result of glucose fermentation. Heterofermentative bacteria ferment glucose to produce lactic, acetic, succinic, and formic acids, as well as CO_2, H_2 and ethanol. Homofermentative bacteria can also ferment glucose to lactic acid by 90%. They lower the pH of the medium (from 7.5) to 5.0 or less.

Citrate utilization may be used to differentiate certain Enterobacteriaceae genus. *E. coli* can't grow on citrate medium; *Enterobacter* grow on the citrate medium. Most members as *Salmonella* utilize citrate, but *Salmonella* Typhi and all members of the genus *Shigella* do not (Table 15.1).

Triple sugar iron agar (TSIA) can differentiate gram-negative microorganisms according to fermentation of glucose, lactose, or sucrose. Acids are produced from sugar, and the medium changes from red to yellow.

Table 15.1 Differentiation of some enteric bacteria by IMViC and HOMoC tests.[a]

Bacteria	I	M	V	C	H	O	M
Escherichia coli	+	+	–	–	–	v	+
Enterobacter aerogenes	–	–	+	+	–	+	+
Pseudomonas aeruginosa	–	–	–		–	–	+
Citrobacter freundii	+	+	–	v	+	+	+
Klebsiella pneumonia	–	v	+	+	–	–	–
Salmonella spp.	–	+	–	v	v	+	+
Salmonella Typhi	–	+	–	–	+	–	+
Shigella spp.	v	+	–	–	–	v	–
Yersinia enterocolitica	v	+	+	–	–	+	+

[a] v = variable, – = negative, + = positive.

In this event, the entire agar turns into yellow color. Anaerobic conditions exist at bottom of the medium in the tube; the rate of glucose utilization occur behind that on the surface of the slant. TSIA is used to determine the ability of a microorganism to produce hydrogen sulfide (H_2S) from sodium thiosulphate. When H_2S production occurs, iron displaces hydrogen and produce ferrous sulfide (FeS), which appears as a black precipitation on the surface area of slant.

A small number of bacteria (such as *Proteus*, *Morganella*, and *Providencia*) produce the enzyme diamine phenylalanase, hydrolyzing amino acids and converting them to phenylpurivic acid. With these features, these three genera are distinguished from other Enterobacteriaceae species.

15.2 Fermentation and oxidation tests

Materials

Microbial cultures: 24 h nutrient broth cultures of *Staphylococcus aureus*, *E. coli*, *Enterobacter aerogenes*, *Proteus vulgaris*, and *Pseudomonas aeruginosa*.

Media: Phenol red glucose broth (PRGB) with inverted Durham tube, nitrate broth with inverted Durham tube, starch agar plate, tryptone soy agar (TSA) plate, Simmons' citrate agar (SCA) slant, phenyl alanine agar (PAA) slant, triple sugar iron agar (TSIA) slant, MR-VP (methyl red-Voges Proskauer) broth and nitrate broth.

Solutions: Gram's iodine solution, methyl red solution, Barritt's reagent A (alpha-naphthol) and B (KOH), nitrite test solution A (contains sulfanilic acid) and B (contains dimethyl-alpha-naphthylamine), powdered pink zinc, oxidase test reagent (1% solution of dimethyl-p-phenylenediamine hydrochloride) and ferric chloride solution (10%, w/v).

Equipments: Sterile test tubes, pipettes (1 mL) in box, loop, needle, water-proof pen, Whatman no: 2 filter paper and sterile Petri dish.

15.2.1 Inoculation

1. Label each broth, agar slants, and plates with the name of media, your initials and name of microorganism which you will inoculate.
2. Inoculate all broths with a loopful labeled culture.
3. Inoculate all plates by streaking with labeled microorganism using the loop.
4. Remove culture using needle, inoculate TSIA slant with labeled microorganism by inserting the needle through the medium down to the bottom of the tube (Fig. 15.2) and before removing the needle from the tube, streak the surface of the slant. Note that this type of inoculation is performed in one operation.
5. Inoculate SCA slant by streaking using a needle from culture.
6. Incubate inverted plates, broths, and slants at optimum growth temperature of labeled microorganisms.

Remember that all tests must always be controlled by comparison with an uninoculated medium.

15.2.2 Evaluations of test

After 24–48 h incubation, place all tubes in a receptacle and Petri plates on desk top.

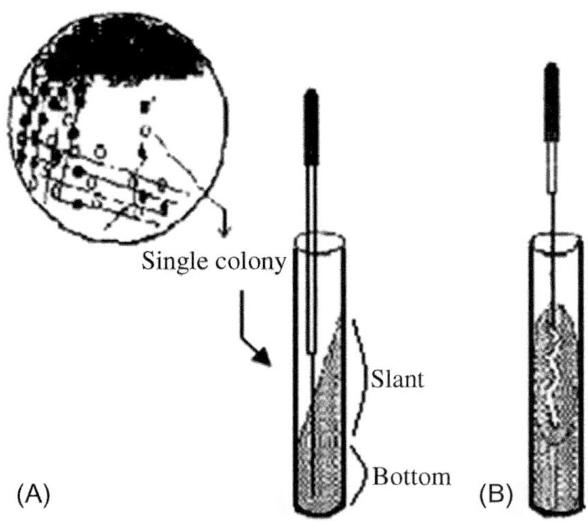

Fig. 15.2 Inoculation of TSIA slant. Pick cells from the center of the selected single colony on plate by needle and transfer it to TSIA slant by (a) stabbing bottom (b) streaking surface of slant under aseptic conditions.

15.2.2.1 *Sugar fermentations*

Durham tube containing PRG broth is used to determine sugar fermentation. If a microorganism can ferment a sugar, acid and/or gas will be produced. The color changes because of the pH indicator in the medium, indicated by indicator in the medium. The acid production leads to color changes on the pH. The phenol red turns to yellow with acid production. Gas production is indicated by the formation of an empty space in the inverted Durham tube which was filled with medium after sterilization. These observations show that bacterium ferments glucose to produce acid and gas. Fig. 15.3a illustrates how this positive tube compares with a negative tube and an uninoculated one. Examine fermentation tubes after 24 h, 48 h, and 5 days. Record results in your laboratory notebook using the following keys:

NR (for *no reaction*): tubes show no color change.

A (for *acidity*): the tube is yellow with no gas, record as acid production.

AG (for *acid and gas*): the tube is yellow and the inverted Durham tube contains gas, record acid and gas production.

(a) (b) (c)

Fig. 15.3 Fermentation tests. (a) Sugar fermentation: Tubes from *left* to *right*: uninoculated, positive (acid with color change-gas collected in Durham tube) and only acid production (color change). (b) MR test: *Tube on left* is positive (color change, acidic pH) and *tube on right* is negative (no color change, neutral pH. (c) VP test: *Tube on left* is negative (no color change) and *tube on right* is positive (color change).

15.2.2.2 Mixed acid fermentation (methyl red test, MRT)

1. After 48 h of incubation, aseptically pipet 2 mL of MR-VP culture into a sterile test tube and label with the name of culture. Set aside this tube containing 2 mL of culture to be used for VP (Voges-Proskauer) test. Repeat this procedure for each culture.
2. Add 3 or 4 drops of methyl red solution into remaining culture in the tube.
3. A color change should appear immediately; if pH of medium is acid, the color of reagent will be reddish (Fig. 15.3b).
4. If the pH of the medium is neutral or basic, color of reagent will be yellow. This indicates negative test.
5. Record results in your laboratory notebook.

15.2.2.3 Butanediol fermentation (VP test)

1. Use 2 mL of MR-VP culture that you reserved from MRT with the name of culture.
2. Add 0.5 mL of Barritt's solution A (alpha-naphthol) and shake the tube side to side.
3. Add 0.5 mL of Barritt's solution B (KOH).
4. Shake tubes vigorously side to side every 20 s during 20 min and wait to aids the complete aeration and development of pink (or red) color.
5. Observing red or pink color from tube indicates a positive VP test (Fig. 15.3c).
6. Record your results in your Laboratory notebook.

15.2.2.4 Oxidase production

Perform oxidase test with plate method or filter paper method.

Plate method. Pour an amount of oxidase test reagent to covert colonies of bacterium on the TSA plate culture. Oxidase positive bacterium colonies first become pink, then change to dark-red and finally black (Fig. 15.4a). Record your results in your laboratory notebook.

Filter paper method. Place the sterile Whatman no-2 filter paper into the sterile Petri dish. Place several drops of oxidase test reagent on the filter paper. Remove a loopful of colony from TSA plate culture and smear the colony over a small area of the paper. The positive color reaction will appear within 10–15 s and oxidase positive bacterium colonies first become pink than black (Fig. 15.4b). Record results in your laboratory notebook.

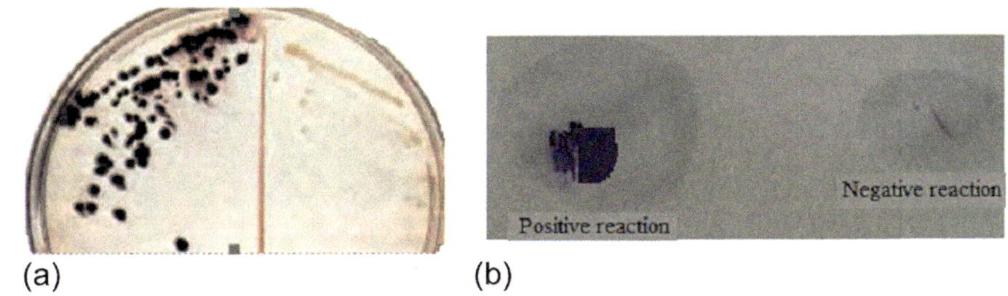

Fig. 15.4 (a) Oxidase Test on Plate culture: oxidase positive colonies with black color on left and negative ones on right. (b) Oxidase Test on filter paper: oxidase positive colonies with black color on left and negative ones on right.

15.2.2.5 TSIA slant reactions

1. Followings and Fig. 15.5 indicate microbial metabolic reactions that can be observed from TSIA slant.
   a. **No color change throughout the medium:** Indicates non sugar fermentation.
   b. **Glucose fermentation**: Retaining red color on slant indicates no fermentation of lactose, yellow bottom indicates glucose fermentation to acid, raising the medium or cracking and furrowing in the medium beside yellow color indicates gas production from glucose.
   c. **Lactose and glucose fermentation**: Yellow slant and yellow bottom indicate lactose and glucose fermentation with acid and gas productions.
   d. **H2S production:** Red slant, black precipitation on slant through inoculation line indicates H_2S production. Yellow butt often obscured by black precipitate.

15.2.2.6 Nitrate reduction

Result of nitrate tests:

(i) Observe the inverted Durham tube in the nitrate broth for the presence of gas (empty area at the top of Durham tube): any trapped gas at the top of Durham tube remove broth and empty space indicates nitrogen gas production (N_2) as on the left side tube (Fig. 15.5a). If there is no gas and no color change, nitrate was not reduced to N_2.

(ii) Reduction of nitrate to nitrite: After incubation, adding 1 mL of sulfanilic acid (solution A) and approximately 1 mL of dimethyl-alpha-naphthylamine (solution B) solutions to each of the nitrate broth cultures and the control. Do not shake the cultures since introducing

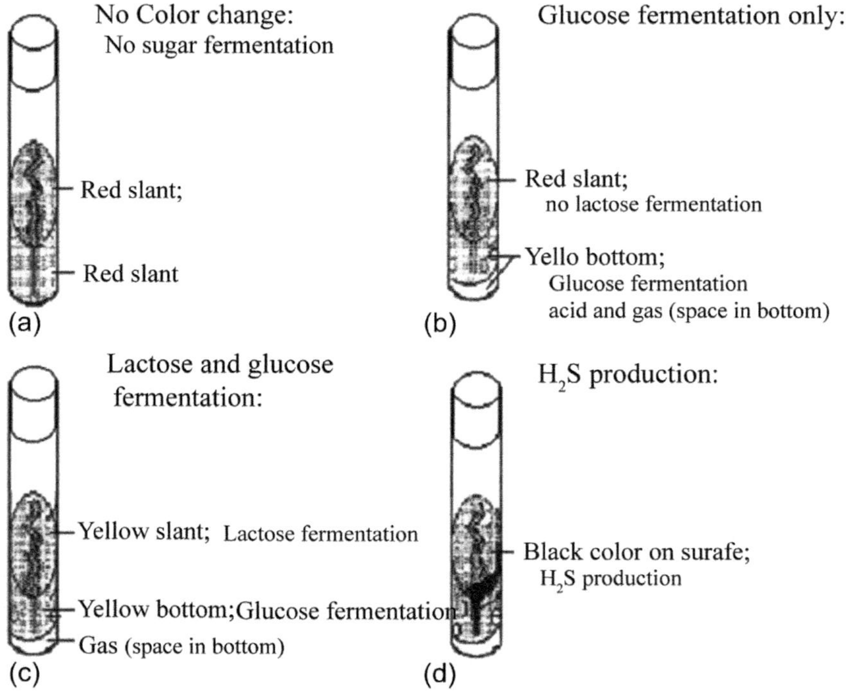

Fig. 15.5 Microbial metabolic reaction results on TSIA slant.

of oxygen interferes with reduction. Appearance of a distinct red color turn brown rapidly, this indicates reduction of NO_3 to NO_2 and the test is positive for nitrate reduction as on the center tube (Fig. 15.5a). No color change indicates negative test.

Caution: Avoid skin contact with dimethyl-alpha-naphthylamine solution is carcinogenic.

(iii) Nitrite reduction or nitrate remains: To the negative broth from activity (ii), add a small amount of powdered pink zinc into the broth in the tube, shake it vigorously, and allow the tube to stand for 10 min. Zinc reduces nitrate to nitrite so a red color appears as on the center tube (Fig. 15.5a), this indicates remain (unalter) nitrate in the broth. This indicates negative nitrate reduction. If no color change occurs upon the addition of zinc as on the right-side tube (Fig. 15.5a), this means that the nitrate has been reduced by the bacterium to nitrite and nitrite to NH_3 or N_2.

(iv) Record results in your laboratory notebook. If results are positive, indicate whether the final product of nitrate reduction is nitrite, ammonia, or nitrogen gas.

15.2.2.7 Phenylalanine deamination

Allow 5–10 drops of 10% ferric chloride solution to flow down over the inoculated PAA slant. To hasten the reaction, use a loop to emulsify the microorganism into solution. The appearance of a deep green color indicates phenylalanine deamination to phenyl-pyruvic acid with the enzyme phenylalanine deaminase as on the left side tube (Fig. 15.5b). No color change indicates non-deamination of phenylalanine as on the right-side tube (Fig. 15.6b). Compare your results with the control medium and record the results on your Laboratory Report.

15.2.2.8 Citrate utilization

1. At the end of incubation period, examine SCA slant for growth, and compare them with control tube for any color change. A change from green to a blue is a positive indication test for the utilization of citrate (as on the left side tube) as a carbon source by microorganism (Fig. 15.6 c). No color change indicates non-citrate utilization as on the right-side tube (Fig. 15.5c).
2. Record results in your laboratory notebook.

(a) (b) (c)

Fig. 15.6 (a) Nitrate reduction test. (b) Phenyl alanine deamination test. (c) Citrate utilization test. For all tests *left* to *right*: positive and negative.

LABORATORY REPORT

Date:
Name:

Lab Title: Oxidation and Fermentation Tests

1. **Draw the growth of bacteria on TSIA slant aerobically and anaerobically. Explain reasons for your drawing?**

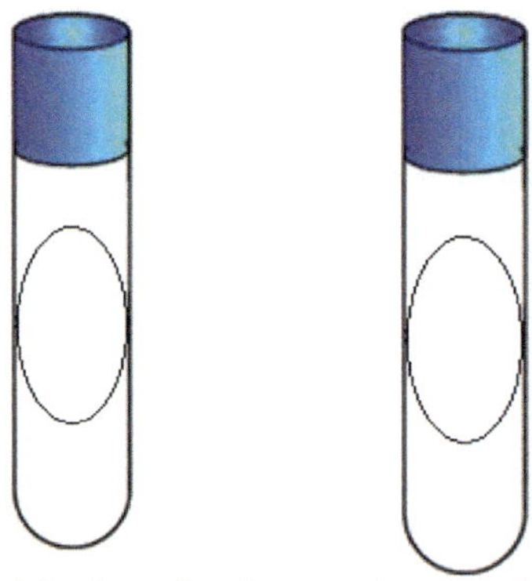

Aerobic incubation Anaerobic incubation

2. **Record your results for the following media.**
   - **(a) Sugar fermentation:**
   - **(b) MR test:**
   - **(c) VP test:**
   - **(d) Oxidase reduction test:**
   - **(e) Nitrate reduction test:**
   - **(f) Phenyl amine deamination:**
   - **(g) Citrate utilization:**
   - **(h) Type of microorganism according to oxygen requirements:**

3. **Explain importance of the hydrolytic tests.**

4. **Discuss possibility of variable reactions in tests.**

5. **Explain reasons of variations of microbial metabolism among bacteria.**

PRACTICE 16

Hydrolytic reactions

16.1 Theoretical information

Many bacteria produce extracellular hydrolytic enzymes, which split complex organic compounds into smaller units in the presence of water (such as hydrolysis of gelatin, starch, casein, fat, tryptophan, and urea). All bacteria can also have endoenzymes functional within cell. Test plays an important role in the identification of bacteria with respect to hydrolysis of organic compounds. Many bacteria and molds (such as *Bacillus subtilis* and *Aspergillus oryzae*) can hydrolyze starch with production of extracellular amylase enzyme. Some microorganisms utilize a particular carbohydrate slowly. For this reason, a tube should be examined both 48 h and 5 days after incubation. Microorganisms are also differing in the utilization of protein as well as carbohydrate. Many different nitrogen-containing compounds are involved in the metabolism of a living cell. Proteins are also hydrolyzed by extracellular proteinases and peptidases. Following hydrolytic reactions are carried out by microorganisms: protein degradations (gelatin), urea hydrolysis, litmus milk reactions, starch hydrolysis, casein hydrolysis, and tryptophan hydrolysis.

16.2 Hydrolytic reactions on different substrates

16.2.1 Litmus milk reactions

Litmus milk medium consists of 10% powdered skim milk and a small amount of litmus. Litmus, when added to rehydrate skim milk, turns colloidal milk suspension from white to purple.

16.2.2 Starch hydrolysis

Many bacteria are able to hydrolyzing starch by extracellular amylases. The starch molecule has two constituents: amylose (a straight chain polymer of 200–300 glucose units) and amylopectin (a larger branched polymer with phosphate groups). Starch hydrolyzing bacteria produce amylases that yield molecules of maltose, glucose, and dextrin.

https://doi.org/10.1016/B978-0-323-91017-0.00008-1

16.2.3 Fat hydrolysis

Microorganisms hydrolyze fats by producing extracellular lipase enzymes. As a result of this hydrolysis, a glycerol molecule and three fatty acid molecules are formed from the fat molecule. Glycerol and fatty acids can be used to synthesize the microbial oils and other cellular components that the microorganism needs for its cellular activities. In many cases, they are oxidized to produce energy in aerobic conditions. Hydrolytic properties of bacteria play a role in the hardness of some foods, such as margarine.

16.2.4 Tryptophan hydrolysis (indole production)

Some bacteria, such as *Escherichia coli*, have the ability to convert the amino acid tryptophan to indole and pyruvic acid by tryptophanase enzyme. In laboratory test, indole can be easily detected with the reagent of Kovacs. This test is used to distinguish *E. coli* from many other enteric bacteria. If the bacterium produces the tryptophanase enzyme, the amino group will be removed from the tryptophan and this produces indole. *E. coli* produces indole, pyruvic acid, and ammonia in exactly 1:1:1 ratio.

16.2.5 Urea hydrolysis

Some gram-negative enteric bacteria can produce urease enzyme. This enzyme hydrolyses urea to ammonia. When microorganisms produce urease enzyme, this enzyme hydrolyzes urea to produce ammonia and raises the pH of medium. As the pH of medium increases, the color of medium changes from a yellow (pH 6.8) to red (pH 8.1 or more) in the presence of phenol red indicator.

$$\underset{\text{Urea}}{O{=}C(NH_2)_2} + H_2O \xrightarrow{\text{Urease}} \underset{\text{Ammonia}}{2NH_3} + CO_2$$

16.2.6 Gelatin hydrolysis

Gelatin is a protein prepared by the hydrolysis of the insoluble protein collagen. The degradation action of a microorganism on an intact protein is analogous to such action on carbohydrates. With a protein, such as gelatin, the enzyme gelatinase accomplishes much the same result yielding simple units called amino acids. The hydrolytic activity of the exoenzyme is manifested by the liquefaction of the nutrient.

16.3 Hydrolytic reaction tests

Materials

Microbial culture: 24 h nutrient broth cultures of *Pseudomonas fluorescens*, *E. coli*, *Alcaligenes viscolactis*, *Proteus vulgaris*, *Enterobacter aerogenes*, and *Staphylacoccus aureus*.

Medium: Starch agar plate, skim milk agar plate, spirit blue agar plate, tryptone broth (1%), nutrient gelatin agar, urea agar slant, and litmus milk medium (10%, pH = 6.8).

Solution: Gram's iodine solution, Kovac's reagent, nitrite test solution A (sulfanilic acid) and B (dimethyl-naphthylamine), and powdered zinc.

Equipments: sterile test tubes, loop, needle and sterile pipettes (1 mL) in box.

16.3.1 Inoculation

1. Label each tube and plates with the name of medium and your initials.
2. Using a water-proof marking pen, divide the plate into six sections. Label each section with the name of microorganism.
3. Inoculate all plates (starch agar and spirit blue agar) by a line streaking.
4. Inoculate all broths (tryptone broth and litmus milk medium), nutrient gelatin stab agar and urea agar slant with labeled microorganism using the loop.
5. Incubate plates (inverted) and tubes at 35°C for 24 h in the incubator.

16.3.2 Hydrolytic reaction evaluations

Perform following evaluations and record results in your laboratory notebook.

(1) Litmus Milk Reactions

Following characteristic reactions might be observed in the litmus milk medium (Fig. 16.1):

- **(i) Alkaline reaction:** A microorganism that can't hydrolyze lactose might well utilize milk proteins as a source of nitrogen, and an alkaline reaction might result. This is indicated by a change in litmus from purple (Fig. 16.1a) to blue (Fig. 16.1b). Many proteolytic bacteria produce acid and basic reactions in the first 48 h of incubation.
- **(ii) Acid curd:** If lactose is fermented to organic acids, a firm, acid curd is formed. As the acidity increases, the curd becomes so solid that there is a syneresis of a clear liquid (whey) from the

Fig. 16.1 Litmus milk reactions: (a) control *(purple)*, (b) alkaline *(blue)*, (c) acid *(pink)*, (d) acid with curd formation; *upper transparent portion* is peptonization; *solid white portion* in bottom is coagulation and litmus reduction; overall redness is interpreted as acid and (e) alkaline with peptonization *(brown top, clear brown bottom)*.

curd. The formation of curd indicates fermentation with the litmus turning from purple (neutral) to red (or pink) as the pH decreases (Fig. 16.1c).

(iii) Litmus reduction. In litmus milk, the litmus serves as an electron acceptor. The pH indicator appears purple, blue, or pink when it is in an oxidized state. Reduction of the litmus by oxidation-reduction (O/R) activities of the bacteria can be shown by loss of the litmus color and it appears white (Fig. 16.1d). This color change usually occurs within 5 days and quite apparent when a pink acid curd begins to turn white.

(iv) Peptonization (or solubilization): It refers to the digestion of milk protein casein by extracellular proteinases (proteolytic enzymes). Peptonization results in the release of large amounts of ammonia; an alkaline condition and the litmus turns a purplish-blue. With further incubation, an opaque clearing of the milk occurs as the casein is hydrolyzed to peptides and amino acids. This opaque (turbid) liquid supernatant often turns into brown (Fig. 16.1e). This reaction can be observed after 48 h, but it is much more obvious after 5 days.

(v) Ropiness. Some microorganisms, such as *Alcaligenes viscolactis,* grow extremely well in litmus milk. Cells become encapsulated because of the rich nutritive environment and from a slimy growth in the medium. This is thick, slimy residue in the bottom of the tube. The "ropiness" can be detected by immersing a sterile loop into the medium. When the loop is removed, a long string of growth becomes obvious.

(2) Starch hydrolysis

After incubation of starch agar plate, carefully flood the surface of the agar with the Gram's iodine solution. If a clear zone appears around a line of bacteria growth, starch hydrolysis was occurred. If the agar immediately adjacent to a line of bacterial growth is dark blue, hydrolyses had not occurred (Fig. 16.2).

(3) Casein hydrolysis

Analyze the skim milk agar streak plates. Presence of a clear zone formation around *B. subtilis* growth indicates casein hydrolysis (Fig. 16.2).

(4) Fat hydrolysis

When the vegetable oil is hydrolyzed in the Spirit blue agar plate by microorganisms, this indicates the decreasing pH sufficiently to produce a dark blue precipitation (Fig. 16.2). Unfortunately, the hydrolytic action of some microorganisms on Spirit blue agar cannot produce a blue precipitate because the pH is not decreased sufficiently.

(5) Tryptophan hydrolysis (indole test)

The liquid medium containing tryptone (1%) is used for this test. About 10 drops of Kovac's reagent is added into tryptone broth culture, gently shake the tube from side to side. In the presence of indole, a deep red color will develop in the alcohol layer and rise to the surface of the medium (Fig. 16.3a). The development of color may take several minutes, and

Fig. 16.2 Hydrolysis on plates: Starch, casein and fat. *(Modified with permission from McGraw Hill; Benson, H.J. Microbiological applications: laboratory manual in general microbiology, 2001, Fig. 49.2, p. 172.)*

Fig. 16.3 (a) Indole test from *left* to *right*: negative and positive and (b) Urease test from *left* to *right*: negative and positive.

repeated shaking may be necessary. Do not consider the test negative until you have observed your tubes for 10–15 min. The development of a red layer on the top of culture is a positive reaction. Add the reagents to the control tube and treat it the same as you did the liquid cultures.

(6) Urea hydrolysis

Examine urea agar slant that is inoculated with *P. vulgaris* (Fig. 16.3b). Presence of a red color in the urea agar slant indicates a positive result. pH indicator (phenol red in the medium) demonstrates production of ammonia by microorganism. Ammonia liberated as a result of urease activity causes a rise in pH from near neutrality to about 8.0. Absence of a red color in the urea agar slant indicates a negative result. The urease negative broth is incubated for a total of 7 days to check for a slow urease production.

(7) Gelatin hydrolysis

Place the tubes of nutrient gelatin into the refrigerator and allow to remain for 15 min. Gelatinase activity is indicated, when the medium remains liquid after refrigeration.

LABORATORY REPORT

Date:

Name:

Lab Title: Hydrolytic Reactions

1. **Report each of hydrolytic reactions, explain reasons of occurrence and not occurrence of these reactions.**

(a) Starch hydrolysis

Bacteria	Type of color change	Reason of result

(b) Casein hydrolysis

Bacteria	Clear zone	Reason of result

(c) Fat hydrolysis

Bacteria	Type of color change	Reason of result

(d) Tryptophan hydrolysis

Bacteria	Type of color change	Reason of result

(e) Urea hydrolysis

Bacteria	Type of color change	Reason of result

(f) Gelatin hydrolysis

Bacteria	Liquid or solid	Reason of result

2. Explain importance ofuse of hydrolytic tests.

3. Explain reasons of variations in hydrolytic tests among bacteria.

4. After inoculating a tube of nutrient gelatin, you incubate it overnight at 35°C. When you remove the tube from the incubator, you notice that the gelatin has liquefied. Does this indicate that the organism produces gelatinase? Why or why not?

SECTION IV

Chemical and physical effects on microorganisms

Any chemical substance or process, kills bacteria, molds, yeasts, viruses, and other microorganisms, is called germicide. The control of microorganisms in the environment or foods generally depends on the creation of conditions that cannot be tolerated by microorganisms. Antimicrobial effects are carried out by microbicidal or macrobiostatic agents. Agents that kill microorganisms are called microbicidal (-cide, kill). Agents that inhibit the growth and survival of microorganisms are called microbiostatic (static, inhibitory), and this effect prevents the growth of microorganisms. Agents that specifically kill (or inhibit) bacteria, fungi and viruses are referred to as bactericidal (or bacteriostatic), fungicidal (or fungistatic) or viricidal (or viristatic), respectively. Control of microorganisms by agents involves the use of antiseptics, disinfectants, antibiotics, high temperature, ultraviolet light, and others. Microorganisms are affected in different ways by their chemical and physical environment. This will perform practices that will determine the effects of some of these chemical and physical factors on microbial growth.

Practices given in this section are as follows:

Practice 18. Antibiotic Sensitivity Test Technique
Practice 19. Effect of Hand Scrubbing on Removal of Microorganisms
Practice 20. Temperature Effects on Microorganisms
Practice 21. Osmotic Pressure and pH Effects on Microorganisms
Practice 22. Ultraviolet Light Effects on Microorganisms
Practice 23. Oxygen Effects on Microorganisms

PRACTICE 17

Antiseptic and disinfectant effects on microorganisms

17.1 Theoretical information

Any substance or process that kills bacteria, viruses, and other microorganisms is called germicide. Depending on the spectrum of antimicrobials killing microorganisms, they are classified as high-, intermediate- and low-level germicides. High-level germicides, such as ethylene oxide, glutaraldehyde, and formaldehyde, can kill all microorganisms and spores. These germicides are used as chemical sterilizer. Intermediate-level germicides (such as 2% solution of glutaraldehyde) kill most microorganisms. The antimicrobial activity of low-level germicides, such as 70% alcohol, is limited to a few types of bacteria, fungi, and viruses. Most antimicrobial agents are low-level germicides, they kill some microorganisms.

Disinfectants (e.g., solutions of phenol, solution of hypochlorite, formaldehyde, alcohols, and halogens) are described as antimicrobial agent used on inanimate objects (such as instruments or surfaces). Antiseptics (such as 70% alcohol and solution of hypochlorite) are usually applied on living tissue. Antiseptics and disinfectants kill microorganisms or inhibit microbial growth within an exposure time. As a skin antiseptic, 70% alcohol is the most widely used antimicrobial. There has been an increasing use of cationic detergents as antiseptics and disinfectants (such as quaternary ammonium compound) and alkyldimethyl-benzylamonium chloride). Quaternary ammonium compound (QAC) is very effective against Gram-negative bacteria, slightly effective against *Mycobacterium tuberculosis,* bacterial spores, and certain viruses (such as hepatitis virus). Effects of antimicrobials on microorganisms are detected by antimicrobial susceptibility techniques. They are Kirby-Bauer technique (disk diffusion technique) and minimum inhibitory concentration technique (broth dilution and agar dilution techniques).

Disk diffusion technique (DDT) is commonly used in the detection of antimicrobial susceptibility testing the many microorganisms. All antimicrobial agents can be tested in DDT. DDT cannot require special equipment. When performed according to the recommendations, it is a reproducible and accurate technique for antimicrobial sensitivity tests (AST). DDT is

Laboratory Practices in Microbiology
https://doi.org/10.1016/B978-0-323-91017-0.00024-X

Table 17.1 Bacterial strains recommended for routine quality control.

Bacteria	Strain number	Bacteria	Strain number
Escherichia coli	ATCC 25922 NCTC 12241	*Enterococcus faecalis*	ATCC 29212 NCTC 12697
Pseudomonas aeruginosa	ATCC 27853 NCTC 12903	*Streptococcus pneumoniae*	ATCC 49619 NCTC 12977
Staphylacoccus aureus	ATCC 29213 NCTC 12973	*Haemophilus influenzae*	NCTC 8468 CIP 54.94

ATCC, American Type Culture Collection, USA; *NCTC*, National Collection of Type Cultures, UK; *CIP*, Collection de Institute Pasteur, France.

used most widely including Mueller-Hinton agar (MHA) with an inoculum corresponding to a McFarland 0.5 turbidity standard. Agar is dispensed in sterile Petri dishes to achieve an even depth of 4.0 ± 0.5 mm that can be provided by pouring 25 and 60 mL agar medium into 90 and 150 mm Petri dish, respectively. A maximum of six disks can be placed on the 90 mm Petri plate and 12 on the 150 mm Petri plate. All inoculum cultures should be used within 15 min and always within 60 min of preparation. The bacterial strains given in Table 17.1 can be used for routine quality control of tests.

The MHA plate surface area must be dried to prevent condensation of water vapor on the lid. Surface area can be dried in an incubator or near Bunsen burner flame or the bottom of plate can be opened quarter over lid and allowed 20–30 min in an incubator. In this technique, a filter paper disk is impregnated with an antiseptic solution and placed on an agar plate inoculated with a microorganism. If the antimicrobial agent inhibits or kills the microorganism, a clear open area of inhibition zone will form around the disk. The magnitude of this region indicates an expression of antimicrobial agent effect on microorganism. In this technique, the relative effectiveness of antimicrobial agents against microorganisms is indicated.

17.2 Antiseptics and disinfectants antimicrobial sensitivity test techniques

17.2.1 Evaluation of antiseptics on skin

Materials

MHA plate, small beaker, 70% ethanol, and 0.1% hypochlorite solutions in the sterile screw-capped bottles.

Procedure

With each antiseptic, perform the steps described in the following procedure:

1. Mark the MHA plate bottom using a waterproof marking pen and divide the plates with two perpendicular lines into four quadrants. Label the quadrants as A, B, C, and D.
2. Perform with unwashed hands. Press your left finger onto the agar surface in the A quadrant.
3. Press your right finger onto the other surface area of agar in the B quadrant.
4. Apply antiseptic to the right and left fingers by dipping the fingers into a beaker of antiseptic for 5 s.
5. Remove your fingers from disinfectant and allow your hands completely dried in air.
6. After drying of your hands, press same left finger onto the C quadrant and right onto the D quadrant.
7. Incubate inverted plates at 35°C for 24 h in the incubator.
8. After incubation, examine the growth on the plate and indicate effects of antiseptic on the microorganisms present on fingers.
9. Record results in your laboratory notebook for the presence or absence of growth and if growth present, indicate the numbers, color, and sizes of colonies.

17.2.2 Antimicrobial susceptibility test by DDT

Materials

Microbial cultures: 24 h Mueller-Hinton broth culture of *Bacillus cereus*, *E. coli* and *S. aureus*.

Media and solution: Prepare antimicrobials using Mueller-Hinton Broth (MHB). MHB and MHA plates, 0.5 McFarland standard, sterile physiological saline solution (PSS), and sterile water.

Antimicrobial solutions in the sterile screw-capped bottles: Zephran solution (1%), phenol solution (5%), formaldehyde solution (8%), benzoic acid solution (0.01%), ethanol solution (70%), and hypochlorite solutions (1.0%, 0.5%, 0.1%, and 0.01%). Prepare antimicrobial solutions using PSS.

Equipment: Sterile filter paper disk (6 mm in diameter Whatman filter paper No: 3) in the Petri dish, sterile screw-capped bottles, forceps, rural, pipette (1 mL), sterile cotton swab, spectrophotometer, and spectrophotometer cuvette (2 pieces).

(a) Preparation of McFarland turbidity standard

In order to standardize the number of microorganisms in the culture for a susceptibility test, barium sulfate ($BaSO_4$) solution can be used as turbidity standard, which can meet the culture turbidity to 0.5 McFarland standard or other turbidity equivalents can be used. $BaSO_4$ 0.5 McFarland standard can be prepared as follows:

1. While constant stirring, add 99.5 mL of 0.18 mol/L H_2SO_4 (1% v/v) into flask containing 0.5 mL of 0.048 mol/L $BaCl_2$ (1.175% w/v $BaCl_2.2H_2O$).
2. After well mixing, the turbidity of standard is adjusted by adding PSS and the absorbance is measured at 625 nm using spectrophotometer. This turbidity of standard refers to a suspension containing about 1 to 2×10^8 colony forming units (cfu)/ml for *E. coli* ATCC 25922.
3. About 4–6 mL $BaSO_4$ solution is transferred into screw-cap tubes.
4. Tightly sealed tubes are stored in the dark at room temperature.
5. Vigorously mixed $BaSO_4$ solution provides a uniform turbidity appearance. If large particles are present or uniform turbidity is not present in the solution, the $BaSO_4$ solution should be replaced.

(b) Quality control for antimicrobial susceptibility test

The control bacterial strain is used to maintain test performance. The control strain is prepared as well as the test bacterium. Control tests should optimally be set up daily. Each new batch of MHA should be tested that all zones are within the acceptable ranges that defined in the EUCAST (European Committee on Antimicrobial Susceptibility Testing) quality control tables.

(c) Preparation of microbial culture inoculum

(i) Preparation from colonies

1. Select several morphologically similar colonies from overnight growth MHA plate.
2. Take the loop from the holder by hand. Sterilize loop by holding it at 90 degree angle through upright over the flame of a Bunsen burner until all the wire becomes redness. Cool the loop for at least 4 s near the flame. Or use a cotton swab.
3. Remove the selected colonies using the sterile loop or the cotton swab.
4. Suspend the colonies in sterile saline PSS in the tube to the density of 0.5 McFarland turbidity.
5. Measure the turbidity of the culture with a spectrophotometric device against 0.5 McFarland turbidity. Alternatively, the turbidity of the suspension can be compared visually to a 0.5 McFarland turbidity standard.
6. If the microbial culture turbidity is not equal to the McFarland turbidity, adjusted to standard by addition of PSS or culture.

7. The culture at standard turbidity should optimally be used within 15 min and always within 60 min of preparation.

(ii) Preparation from broth

1. Take the loop from the holder by hand. Sterilize loop by holding it at 90 degree angle through upright over the flame of a Bunsen burner until all the wire becomes redness. The entire wire must be heated and cool the loop for at least 4 s near flame.
2. Inoculate one loopful 24 h nutrient broth culture into MHB.
3. Incubate the inoculated MHB at 35°C in the incubator for 24 h.
4. Place the MHB culture into centrifuge tube and centrifuge at 2500 g for 10 min.
5. Remove the supernatant (which contains mainly water) with decantation or pipetting.
6. Resuspend the pellet in 2 mL of PSS. Break down the pellet by scraping and by vertexing the tube.
7. Repeat centrifugation at least once and discard the supernatant which mainly contains water.
8. Resuspend the pellet in 2 mL of PSS. If the pellet remained turbid, repeat centrifugation.
9. Microbial cells are mixed with PSS and continue to add PSS until to achieve McFarland turbidity.
10. Measure the turbidity of the culture with a spectrophotometric device that has been calibrated with a McFarland standard according to the manufacturer's instructions. Alternatively, the turbidity of the suspension can be compared visually with 0.5 McFarland turbidity.
11. The culture at standard turbidity should optimally be used within 15 min and always within 60 min of preparation.

17.2.2.1 Disk diffusion technique

With each culture provided, perform the steps described in the following procedure and Fig. 17.1.

1. With a waterproof marking pen, label each MHA plate with the name of microorganism and your initial.
2. Mark the bottom of the Petri plate with lines that divide it into six sections. Label each section with an antiseptic agent and its concentration.
3. After adjusting the turbidity of the inoculated culture, dip a sterile cotton swab into this culture. The cotton swab is rotated several times in culture and excess liquid is removed above the liquid level by pressing the swab firmly against the inner wall of the tube. This removes excess microorganism and water from the stick.

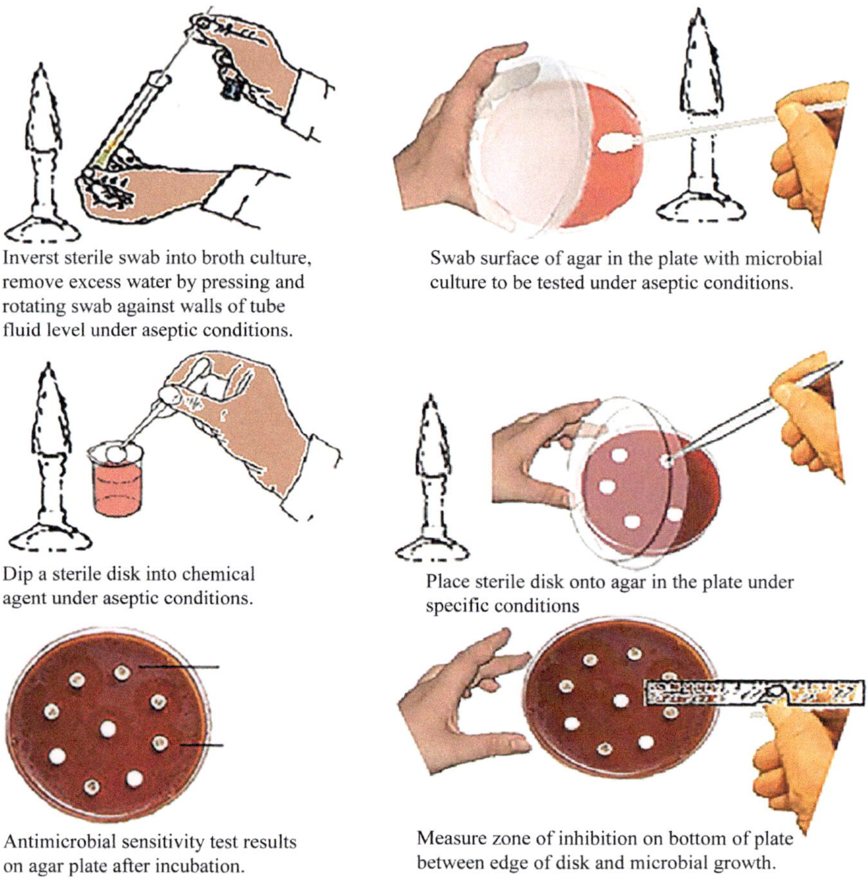

Fig. 17.1 Disk diffusion technique for evaluating an antimicrobial effect.

4. The dried MHA plate is inoculated by swabbing the surface area of agar. This procedure is repeated by swabbing with rotating the plate approximately 60 degree each time to ensure an even distribution of inoculum.
5. The lid may be left open for 20–30 min near Bunsen burner flame to absorb or evaporate any excess surface moisture. Lid of the plate must be free from water and water vapor.
6. Take forceps, deep into alcohol in the bottle, flame the forceps to burn alcohol and cool near flame.
7. Pick up a sterile disk with forceps and dip the disk into the antiseptic solution in the bottle. The disk should be touched several times on the inside sterile beaker wall to remove excess fluid.

8. Place disk to the labeled place on the surface of the agar plate at least 15 mm from edge of plate. To secure the disk firmly contact to the medium, apply light pressure on the disk on the agar with tip of the same forceps.
9. Repeat this procedure for each antimicrobial agent steps 7 and 8. The disks must be distributed evenly so that they are not closer than 24 mm from center to center of disk. Ordinarily, not more than six disks is placed per plate.
10. Incubate inverted Petri plates in the incubator at 35°C for 16–20 h within 15 min after replacing disks onto agar.
11. After incubation, examine each plate and look for the presence of growth-inhibition zones. A correct inoculum and satisfactorily streaked plates should result in an even confluent lawn (complete clear area) of growth. If individual colonies can be seen, the inoculum is too light and the test should be repeated.
12. Using the rural, zones of inhibition are measured to the nearest whole millimeter from the edge of the disk to edge of the growth (Fig. 17.1). Petri plate is illuminated with reflected light by being held 30 cm from the eye in front of a black and non-reflective background. It should be measured on the bottom of the plate, not on the surface of the medium or not on the lid of plate. Measure antimicrobial zone least three times from different plates.
13. Record the results in your laboratory notebook indicating the zone size in mm.

17.2.2.2 Dilution techniques

Dilution techniques are widely used to determine the susceptibility of microorganisms to antimicrobial agents. There are two dilution techniques in the prediction of minimum inhibitory concentration (MIC) of antimicrobial agent on microorganism: (a) broth dilution technique and (b) agar dilution technique. The advantage of the dilution technique that the same tubes can also be used for the detection of minimum bactericidal concentration (MBC) and maximum tolerable concentration (MTC).

(1) Broth dilution technique

Materials

Microbial cultures: 24 h MHB cultures of *S. aureus*, *Klebsiella pneumonia* and *E. coli*. 0.5 McFarland standard.

Media and solutions: MHB (200 mL) and sterile distilled water.

Antimicrobial solutions in the sterile screw-capped bottles: Zephran solution (1%), phenol solution (5%), formaldehyde solution (8%), benzoic acid solution (0.01%), ethanol solution (70%), hypochlorite solutions (1.0%, 0.5%, 0.1%, and 0.01%). Prepare amicrobial solutions using PSS.

Equipment: Sterile pipettes (10 and 1 mL) in boxes, loop, vortex, sterile screw-capped tubes (7.5 × 1.3 cm), receptacle.

(a) Minimum inhibitory concentration (MIC)

1. Label the sterile screw-capped tubes for the antimicrobial concentrations.
2. Place two rows of 12 sterile 7.5 × 1.3 cm capped tubes in the receptacle.
3. Prepare microbial inoculum at McFarland standard as explained in Section 17.2.2c.
4. Using the pipette, add 4 mL of sterile MHB into all sterile tubes.
5. Mix the contents of the stock antimicrobial agent by vertexing. Transfer 2 mL of antimicrobial agent into the first tube in each row using pipette and mix the contents by vertexing.
6. Transfer 2 mL into the second tube from the first tube in each row.
7. Continue preparing dilutions of antimicrobial agent in this way by transferring 2 mL to other tubes as many as dilutions (up to 10th tubes) are obtained. Remove 2 mL of broth from each of 10 tubes. In each case, use new sterile pipette and mix the contents by vertexing after each transfer.
8. Place 2 mL of stock antimicrobial into tubes in the 11th row as antimicrobial control and last row is used as medium control.
9. Inoculate tubes from 1st to 10th tubes with one drop (0.1 mL) of a culture at McFarland turbidity.
10. Incubate the tubes at 35°C for 18 h in the incubator.
11. Calculate the concentration of antimicrobial agents per ml of medium in tubes.
12. After incubation, look for the presence or absence of growth (turbidity) in tubes. Determine MIC values of antimicrobial agents for each microorganism. MIC is expressed as the lowest dilution, which inhibits growth of microorganism by lack of turbidity in the tube.
13. In control tubes, growth must not be observed.
14. The test must include a second set of the same dilutions inoculated with the standard microbial strain. The purpose of the control is to confirm by its MIC that the antimicrobial level is correct.
15. Use the tubes with the lack of turbidity for MBC test and with the last growth for MTC test.

(b) Minimum bactericidal concentration (MBC)

1. Note the lowest concentration of antimicrobial inhibiting growth of microorganism (no turbidity) and record this as the MIC.
2. Subculture all tubes not showing visible growth starting from MIC by streaking a loopful culture over a quarter of MHA plate.
3. Incubate inverted plates at 35°C for 24 h in the incubator.
4. Look at for growth of microorganisms (colony formation) on the Petri plate.
5. Concentration of antimicrobial on tubes with the highest dilution without growth on agar medium is taken as MBC.

(c) Maximum tolerable concentration (MTC)

1. Note the highest concentration of antimicrobial agent that has no effect on the growth of test microorganism (turbidity).
2. Record this tube as an MTC.

(2) Agar dilution technique

Agar dilution techniques for the detection of MIC value have an advantage that it is possible to test several microorganisms on the same plate. But it cannot available to indicate MBC and MTC.

Materials

Microbial culture: 24 h MHB cultures of *S. aureus*, *K. pneumonia*, and *E. coli*.

Media and solutions: MHA stab in tubes (24 mL), 0.5 McFarland standard and sterile distilled water.

Antimicrobial solutions in the sterile screw-capped bottles: Zephran solution (1%), phenol solution (5%), formaldehyde solution (8%), benzoic acid solution (0.01%), ethanol solution (70%), and hypochlorite solutions (1.0%, 0.5%, 0.1%, and 0.01%). Prepare amicrobial solutions using PSS.

Equipment: Sterile pipettes (10 and 1 mL) in box and loop.

Procedure

1. Place the MHA stabs in the boiling water bath (at 60°C).
2. Label the bottom of the sterile Petri dishes with antimicrobial concentration and your initials using waterproof marking pen.
3. Prepare microbial inoculum at McFarland standard as explained in Section 17.2.2c.
4. Add 1 mL of diluted antimicrobial solution into 24 mL of MHA at not more than 55°C and mix well.
5. Pour mixture into a sterile Petri dish (90 mm in diameter) to provide 4.0 ± 0.5 mm thickness and mix thoroughly.

6. The factor of agar dilution must be considered in the calculation of antimicrobial concentration.
7. After the agar is solidified in plate, they should be well dried at 35°C with their lids tipped in inverted plate for 20–30 min in an incubator without air circulation.
8. Mark bottom of Petri plate with 5 mm diameter circle 20 mm apart from each other.
9. Labeled each circle with a microbial name and inoculate with culture (at McFarland turbidity) with loop (0.001 mL) and spread over a 5-mm circle area. The inoculum must be dried on the plate.
10. Incubate inverted plates at 35°C for 16–20 h in the incubator.
11. The antibiotic concentration of the first plate showing inhibition (no growth) is taken as the MIC for the microorganism.
12. Control broth tubes and standard microbial strain must be used to standardize the results.

17.2.2.3 Agar well diffusion technique

Materials

Microbial culture: 24 h Mueller-Hinton broth cultures of *S. aureus*, *K. pneumonia*, and *E. coli*.

Media and solutions: MHA plates and sterile distilled water.

Antimicrobial solutions in the sterile screw-capped bottles: Zephran solution (1%), phenol solution (5%), formaldehyde solution (8%), benzoic acid solution (0.01%), ethanol solution (70%), hypochlorite solutions (1.0%, 0.5%, 0.1%, and 0.01%). Prepare amicrobial solutions using PSS.

Equipment: Sterile pipettes (10 and 1 mL) in box and loop.

Procedure

1. Prepare microbial inoculum turbidity at McFarland standard as explained in Section 17.2.2c.
2. Dip a sterile cotton swab into inoculum culture within 15 min after preparation of turbidity. The cotton swab is rotated several times in culture and excess liquid is removed above the liquid level by pressing the swab firmly against the inner wall of the tube. This removes the excess microorganisms and water from the swab.
3. The dried surface of a MHA plate is inoculated by swabbing the surface area of agar. This procedure is repeated by swabbing in three directions with rotating the plate approximately 60 degree each time to ensure an even distribution of inoculum.

4. The lid may be left open for 20–30 min near Bunsen burner flame to absorb (and evaporate) any excess surface moisture.
5. Mark bottom of Petri plate with 4 mm diameter 20 mm apart from each other and label with an antimicrobial solution.
6. Wells were open in the agar for 4 mm diameter with an agar piercing (cork borer).
7. Fifty 50 μL of an antimicrobial solution is dispensed into the agar wells.
8. Allow the plate to dry the antimicrobial solution in the agar well.
9. Control broth tubes and standard microbial strain must be used to standardize the results.
10. Inverted plates are incubated at 35°C for 16–20 h in the incubator.
11. After incubation, the inhibition zone is measured in mm from the edge of the agar well up to growth edge of microorganism. Three inhibition zone measurements were taken for each well and averaged. Reference commercial disks are used.

LABORATORY REPORT

Date:
Name:

Lab Title: Antiseptics and Disinfectants Effects on Microorganisms

1. **Record the results for antiseptic effects on microorganisms on the skin**

	Application	**Colony morphology**		
		Number	**Colors**	**Sizes**
		70% Ethanol solution		
Before wash	Left finger			
	Right finger			
After wash	Left finger			
	Right finger			
		0.1% Hypochlorite solution		
Before wash	Left finger			
	Right finger			
After wash	Left finger			
	Right finger			

2. **Record the antimicrobial susceptibility disk diffusion test results for all microorganisms and antimicrobials.**

Antimicrobial	**Diameter inhibition zone (in mm)**											
	B. cereus				***E. coli***				***S. aureus***			
	1st	**2nd**	**3rd**	**Aver.**	**1st**	**2nd**	**3rd**	**Aver.**	**1st**	**2nd**	**3rd**	**Aver.**

3. Record the antimicrobial susceptibility test results of broth dilution technique for all microorganisms and antimicrobials.

	B. cereus			*E. coli*			*S. aureus*		
Antimicrobial	**MIC**	**MBC**	**MTC**	**MIC**	**MBC**	**MTC**	**MIC**	**MBC**	**MTC**

4. Record the antimicrobial susceptibility test results of agar dilution technique for all microorganisms and antimicrobials.

	Diameter inhibition zone (in mm)											
Antimicrobial	***B. cereus***				***E. coli***				***S. aureus***			

5. **Record the antimicrobial susceptibility test results of agar diffusion well technique for all microorganisms and antimicrobials.**

	B. cereus	***E. coli***	***S. aureus***
Antimicrobial	MIC	MIC	MIC

6. **Use the data above to indicate the most effective agent and the least effective agent on each microorganism. Discuss reasons of effectivity and effectiveness.**

7. **Discuss the efficiency of antiseptics on hands.**

8. **What conclusions can be derived from these results?**

9. **Differentiate antiseptic and disinfectant. Give examples for each.**

10. **Which factors can influence the size of zone of inhibition produced by a chemical agent?**

11. **How might physical differences between Gram-positive and Gram-negative bacteria contribute to differences in chemical resistances?**

12. **Explain importance of antimicrobial tests performing in food industry.**

13. **Why do we classify chemicals as disinfectant or antiseptics and not as chemical sterilant?**

14. In general, what effect does alcohol have on the level of skin contaminants?

15. What do the colonies present within zone of inhibition indicate?

PRACTICE 18

Antibiotic sensitivity test technique

18.1 Theoretical information

Antibiotics (anti: against, bios: life) are therapeutically useful group of chemical compounds. The production of antibiotics by microorganisms is a common occurrence. Antibiotics can be described as a low-molecular weight metabolic byproduct of a microorganism that can kill or inhibit other microorganisms. Many of the lactic acid bacteria produce bacteriocins (such as *Lactobacillus lactis* subsp. *lactis* produces nisin) as the antimicrobial agents. Drugs, on the other hand, are antimicrobial agents that are man-made.

The antimicrobial activity of an antibiotic on microorganism can be tested using different technique to determine the minimum amount required to inhibit a microorganism. The most common technique used to detect the antimicrobial effect of the antibiotic is the disk diffusion technique, known as the Kirby-Bauer technique. The Kirby-Bauer technique is a standard technique that considers all variables. The medium recommended for use in this technique is Mueller-Hinton agar. The pH of this medium should be adjusted to 7.3 and sterile medium should be poured homogeneously into sterile Petri dishes with 4 mm thick. This requires pouring 60 mL into a sterile 150 mm Petri dish and 25 mL into a sterile 90 mm Petri dish (90 × 17 mm). The inoculation of the microorganism on the surface of agar medium in the Petri plate is carried out with a sterile cotton swab from the liquid culture at the McFarland turbidity. After 16–20 h of incubation, the Petri plate is examined on a dark place for the diameter of the clear zone. The antimicrobial effects of antibiotics are defined by the size of the inhibition zone (clear area) of the microorganism formed around the antibiotic discs used. The diameter of these regions formed on the agar surface can be measured precisely with a ruler nearest millimeter. The inhibition zone varies according to the type of the antibiotic, the size of the inoculum, the ambient conditions, and many other factors. In antibiotic sensitivity test, microbial culture must be used in McFarland turbidity. This turbidity is provided by comparing the turbidity of culture with 0.5 McFarland turbidity solution. This turbidity is prepared from $BaSO_4$.

Laboratory Practices in Microbiology
https://doi.org/10.1016/B978-0-323-91017-0.00007-X

Differences of antibiotics from antiseptics and disinfectants: (i) antibiotics can be biosynthesized, (ii) antibiotics travel in the blood stream to all parts of the body, but never antiseptics or disinfectants are found in the body and (iii) antibiotics have low toxic and destruction effects on the body cells while bactericidal on microorganisms.

18.2 Kirby-Bauer antibiotic sensitivity test technique

Materials

Microbial cultures. 24 h Muller-Hinton broth cultures of *Staphylococcus aureus*, *Escherichia coli*, *Pseudomonas aeruginosa*, and *Klebsiella pneumoniae*.

Medium and antibiotics. Mueller-Hinton agar (MHA) plates (150 mm in diameter) and individual antibiotic discs in separate container.

Equipments. Rural, forceps, sterile pipettes (1 mL) box and standard McFarland solution.

Procedure

With each culture provided, perform the steps described in the following procedure and Fig. 18.1. Refer the procedure in Practice 17 for the preparation of 0.5 McFarland standard turbidity.

1. Using a water-proof marking pen, label each plate with the name of microorganism and your initial, and divide the plate into six sections. Ordinarily, no more than six discs on the plate are used.
2. Inoculate one loopful culture from MHB into MHB.
3. Incubate the broth in an incubator at 35°C until the appearance of the turbidity equivalent to the 0.5 McFarland standard (usually within 2–6 h).
4. If turbidity higher than the standard, broth culture turbidity may be adjusted with sterile saline or MHB to the 0.5 McFarland turbidity (from 0.008 to 0.1 absorbance at 652 nm). To perform this step properly, either a spectrophotometric device can be used or if done visually, adequate light is needed to visually compare the inoculum tube.
5. After adjusting the culture turbidity, a sterile cotton swab is dipped into the culture within 15 min. The swab should be rotated several times and pressed firmly on the inside wall of the tube above the fluid level. This will remove excess inoculum from the swab.
6. Inoculate the dried surface of the agar medium with the cotton swab by streaking the swab over the entire sterile agar surface. This procedure is repeated by streaking more times and rotating the plate approximately 60 degree each time to ensure an even distribution of inoculum.

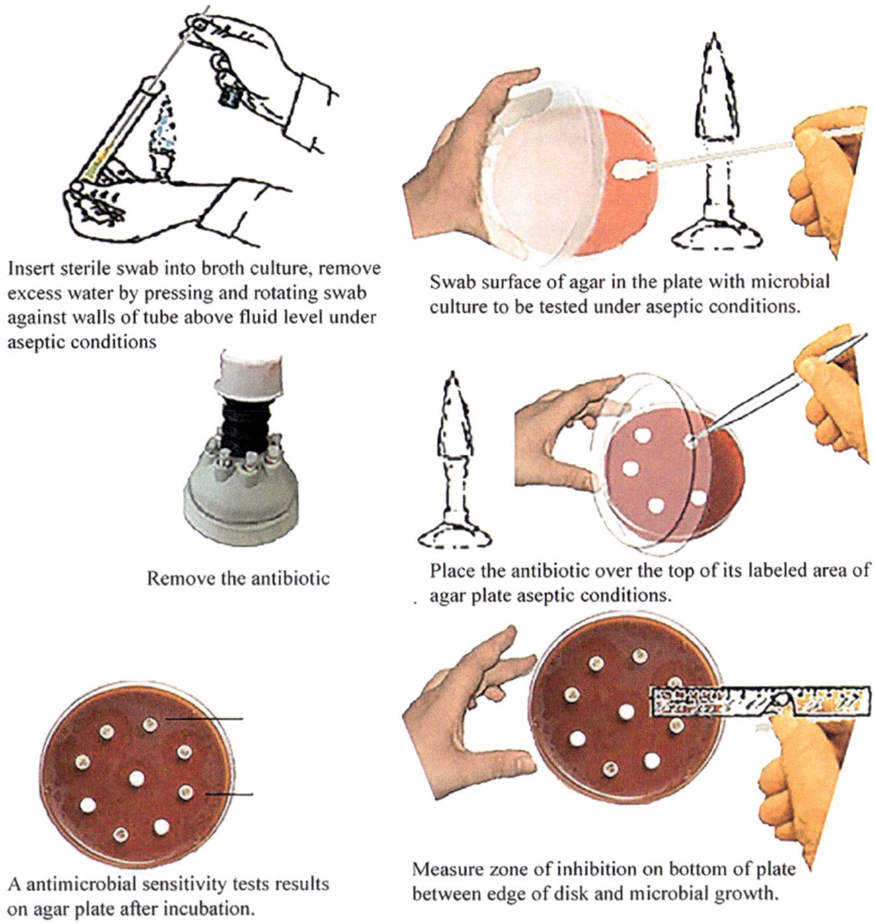

Fig. 18.1 Kirby-Bauer antibiotic sensitivity technique.

Cover the surface of the agar evenly swabbing. A final sweep should be made on the agar rim with the swab.

7. The lid may be left open for 20–30 min near Bunsen burner flame to allow for any excess surface moisture to absorb (and evaporate) before applying the antibiotic discs.
8. Remove the antibiotic plunger and place the plunger over the top of the labeled agar surface on plate. Firmly and quickly push down on the plunger to release the discs onto the agar. Or pick up an antibiotic disc with flamed forceps, place disc to the labeled area on the surface of the agar plate.

9. Keep each disc at least 15 mm from edge of plate. Each disc must be pressed down to ensure complete contact with the agar surface. Repeat this step for each antibiotic disc.
10. Antibiotic discs must be placed onto plates with no closer than 24 mm from center to center of discs.
11. Incubate inverted plates in an incubator at 35°C for 16–20 h within 15 min after discs are placed.
12. After incubation, each plate is examined for the growth inhibition zone presence. The diameters of the zones of complete inhibition (as judged by the unaided eye) are measured from the edge of the disc to edge of the growth (Fig. 18.1). Zones are measured to the nearest whole millimeter (±1 mm) using a ruler, which is held on the back of the inverted Petri plate. Measure the zone three times from different sides. The Petri plate is held a few cm above a black, nonreflecting background and illuminated with reflected light. It should not be measure on the surface of the medium or on the lid of plate.
13. If you observe individual colonies on agar medium, this test must be repeated.
14. Record the results in your laboratory notebook.

LABORATORY REPORT

Date:

Name:

Lab Title: Antibiotic Sensitivity Test Technique

1. Record the results for tested microorganisms.

	Inhibition zone diameter (in mm)											
	B. cereus				*E. coli*				*S. aureus*			
Antibiotic	1	2	3	Average	1	2	3	Average	1	2	3	Average

2. Use the data above to indicate the most effective agent and the least effective agent on microorganisms.

3. Why might Gram-positive and Gram-negative respond differently to the same antibiotic?

4. Explain importance of the use of antibiotic sensitivity tests on microorganisms in food industry.

5. How do the organisms acquire drug resistance?

6. What factors other than the agent being tested can affect the zone of inhibition?

7. How do you differentiate between sensitive and resistant organisms in antibiotic sensitivity test?

8. What do the colonies present within zone of inhibition indicate?

PRACTICE 19

Effect of hand scrubbing on removal of microorganisms

19.1 Theoretical information

Hand disinfection is important in preventing the spread of microbial diseases. Washing hands before touching any food and rubbing hands before producing food should be a common application. Hand will be washed with soap and washing will remove the temporary and settled microorganisms from the surface of hand. Microorganisms firmly placed on the hands can be removed slowly by washing and rubbing the hands. Microorganisms on the surface of the hands mostly consist of low pathogenic staphylococci. They are less sensitive to the effect of antiseptics than other microorganisms. In hand scrubbing, the effectiveness of the time elapsed in the removal of microorganisms from the hands will be evaluated. In order to determine the effect of brushing in reducing the bacterial flora from the hands, washing will be carried out in the sterile basin using sterile water. Later, in the samples to be taken from this water, the number of microorganisms will be determined using plate method.

19.2 Hand scrubbing

Materials

Media and solutions. Brain heard infusion agar (BHIA), alcohol and sterile 9 mL diluent (0.1% peptone water).

Equipments. Sterile pipettes (1 mL) in box, L-shaped glass spreader, sterile scrub brushes and 5 basins (or 2000 mL beakers) containing 1000 mL of sterile water.

19.2.1 Hands brushing and scrubbing

The basic hand washing steps are shown in Fig. 19.1 are detailed below.

1. Before starting to scrub the hands, both students should carefully read the entire procedure and envision what to do.
2. The other students should be divided into five groups (A, B, C, D, and E).

https://doi.org/10.1016/B978-0-323-91017-0.00029-9

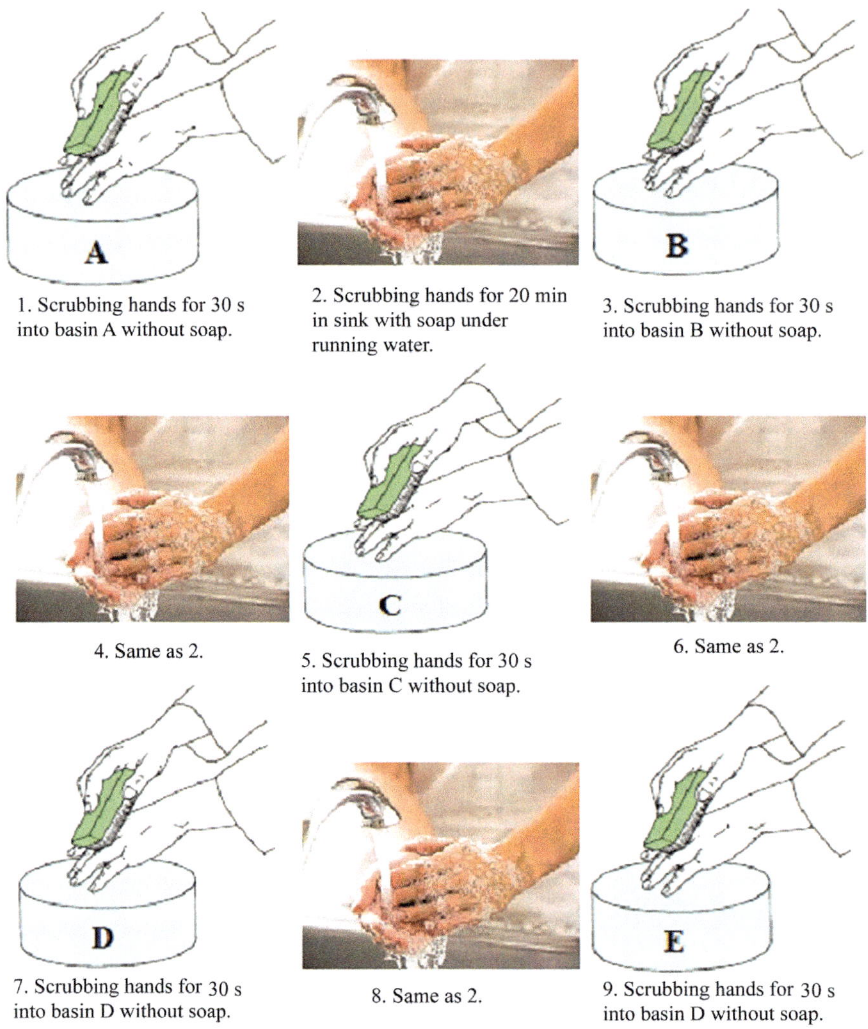

Fig. 19.1 Hand scrubbing in basins. *(Modified with permission from McGraw Hill; Benson, H.J. Microbiological applications: laboratory manual in general microbiology, 2001, Fig. 44.1, p. 149.)*

3. In this experiment, two students should be indicated to brush their hands in each group, and a brusher should be placed in the classroom, where the student can easily wash their hands.
4. While one of the students brushing his hands in water, the other student should help to read and follow the procedure, and provide the necessary material.

5. The procedure should be carried out in parallel.

(a) Hand brushing

Step 1. The entire surface area of each hand should be rubbed into the basin A with a sterile brush for 30 s. At this stage, soap will not be used for washing hands. The successful performance of this step depends on the level of implementation of the following processes:

- The same time should be used to wash each hand (30 s),
- While washing hands, the same level of activity should be maintained in each hand and.
- Under nails and surfaces should be rubbed thoroughly.

After the scrub of the hands is completed, the students in the Group A should be informed for removing sample from scrubbed water.

Step 2. Using the same brush as above, scrub with soap for 2 min using running tap water to moisten in a sink and rinse hands. In this washing, each hand is reserved for 1 min. After completing the scrub, both hands should be rinsed in tap water for 5 s at the same time. Brush should be discarded.

Step 3. With a new sterile brush, hands should be brushed into basin B, similar to step 1. However, soap will not be used in this brushing process. The students in the Group B should be informed for removing sample from scrubbed water.

These procedures should be applied in the steps 5, 7, and 9 of Fig. 19.1, using basins C, D, and E. Students in these groups should also be removed sample after completing their hands scrubbing.

Remember. Fresh sterile water must be used for each of these basins.

After scrubbing. After all scrubbing has been completed, the scrubbed hands should be dried and apply hand lotion.

(b) Microbial count by spread plate technique

During application of the scrub, the instructor will divide the class into five groups (Basins A, B, C, D, and E). Each group will carry out inoculation into six plates (by dilution up to 10^{-5}) for each of the five basins per student hand washing. The bacterial number per milliliter for each basin must be counted by these groups.

1. Label the BHIA plates with dilution number, scrubbing letter, and your initials.
2. As soon as the scrubber has prepared in your basin and perform inoculations with the following procedure:
    - Using an L-shaped stirring rod, stir the water in basin for 15 s under aseptic conditions. Stirring rod must be sterilized before using by

immersing in alcohol and flaming through Bunsen burner flame. For accurate results, all groups must use same stirring procedure.

- Remove 1 mL of water from the basin and add into sterile 9 mL diluent (0.1% peptone water) with a sterile pipette and prepare necessary dilutions up to 10^{-5}.
- Spread plate nondiluted/diluted sample on surface of agar medium using L-shaped glass spreader.
- Incubate inverted plates in an incubator at 35°C for 24 h.

3. After incubation, select the plates containing the best colony distribution between 30 and 300 colonies. Count the number of colonies on the plates and record results on your laboratory report chart. Calculate number of microorganisms (either bacteria or fungi) per ml of brushed water.

LABORATORY REPORT

Date:

Name:

Lab Title: Effect of Hand Scrubbing on Removal Microorganisms

(1) Record average counting results (cfu/ml) onto following tables.

Hand scrub	Dilutions					
	None	10^{-1}	10^{-2}	10^{-3}	10^{-4}	10^{-5}
A						
B						
C						
D						
E						

(2) What conclusions can be derived from this exercise?

(3) What might be an explanation of a higher count in Basin B than in D?

(4) Why is it so important that food producers scrub their hands prior to work on food production even though they wear rubber gloves?

(5) Explain importance of microorganism in the removing of microorganisms from hand?

(6) What environment is your microbe often found in?

PRACTICE 20

Temperature effects on microorganisms

20.1 Theoretical information

Metabolic activities of microorganisms are carried out through enzymes. There are many factors affecting the growth of microorganisms as well as the enzyme activities, the most important of which is temperature. There are minimum, optimum, and maximum growth temperatures that affect the metabolic activities of microorganisms. At the optimum temperature, the activities of enzymes and growth of microorganisms are maximum. Microbial growth is inactivated near minimum temperature. Since enzymes are inactivated near and below the minimum temperature. Growth of microorganisms decreases from optimum to maximum and enzyme denaturation start at near maximum temperature. At the maximum temperature, an enzyme becomes denature. Microorganisms grow in different temperature ranges from about 0°C to over 90°C. Microorganisms can be divided into three groups with respect to their growth temperature requirements (Fig. 20.1). Psychrophiles that can grow between −5 and 20°C with optimum from 12 to 15°C, mesophiles that grow between 16 and 45°C with optimum from 30 to 40°C, and thermophiles that grow at high temperatures between 45 and 70°C (not below 35°C) with optimum from 55 to 60°C. The mesophiles are further subdivided into psychrotrophic and thermoduric. Psychrotrophes grow with optimum 20°C and growth temperature ranging from −5 to 30°C. Thermodurics will grow with optimum between 50 and 55°C and growth temperature ranging from 20 to 65°C. In attempting to compare the susceptibility of different microorganisms to elevated temperatures, two terms are used: the thermal death point (TDP) and the thermal death time (TDT). TDP is the temperature at which a microorganism is killed in 10 min. TDT is time required to kill a microorganism at a given temperature. Different factors (such as pH, moisture, composition of medium and age of cells) influence temperature effects on microorganisms. In this practice, the temperature effecting microorganisms are evaluated.

Laboratory Practices in Microbiology
https://doi.org/10.1016/B978-0-323-91017-0.00025-1

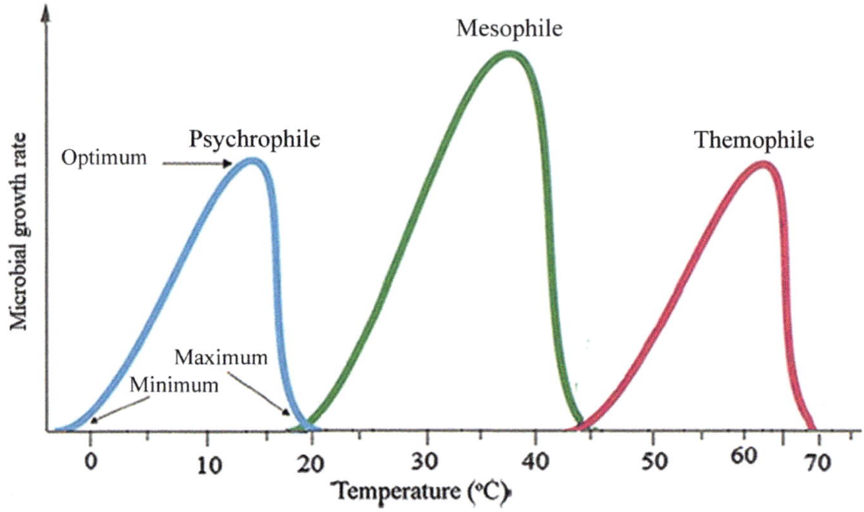

Fig. 20.1 Graph of bacterial growth rate as a function of temperature.

20.2 Indication of temperature effects on microorganisms

Materials

1. 24 h nutrient broth cultures of *Escherichia coli*, Klebsiella pneumonia, Bacillus subtilis, and Yersinia enterocolitica.
2. Nutrient agar plate, nutrient agar (NA) slant and nutrient broth (NB) in tubes.
3. Water bath, thermometer, loop, pipettes (1 mL) in box, glass-spreaders, alcohol in beaker, incubator (at 5, 25, 35, and 45°C).

20.2.1 Temperature effects on growth of microorganisms

1. Label a slant and a broth with each of 5, 20 35, 45, and 50°C for each microorganism. Also label them with the name of bacteria and your initials.
2. Take the loop or needle from receptacle by hand. Sterilize loop by holding it at 90 degree angle through upright over the flame of the Bunsen burner until all the wire becomes redness. The entire wire must be heated and cool the loop for at least 4 s near flame.
3. Using free hand, take labeled liquid culture from receptacle and gently shake the tube to disperse the culture.
4. Remove one loopful of liquid culture and inoculate slant by streaking.
5. Inoculate broth with same culture.
6. Carry out inoculation with other cultures as indicated in 3–5.

7. Incubate the slants and broths at the labeled temperature from 24 to 48 h.
8. Evaluation of results after incubation:

Slant culture: Compare the nutrient agar slant cultures with respect to amount of growth and pigment formation, and record the results in your laboratory notebook.

Broth culture: The degree of turbidity in broth cultures after shaking should be compared and recorded. Tubes without turbidity should be decided by comparison with the control broth.

Record growth of microorganisms from slant and broth on the laboratory report in the following manner: No growth = none, least growth = +, more growth = ++ and most growth = +++. Indicate minimum, optimum, and maximum growth temperature for each of tested bacteria.

20.2.2 Temperature lethal effects

The overall procedure for the temperature lethal effects on microorganisms is illustrated in Fig. 20.2.

1. Label plates with holding time (10, 20, 30, and 40 min), your initials and the temperature at the bottom.
2. Each group is assigned with one culture of microorganisms to be heated at a specific temperature (control, 40, 50, 60, 70, and 80°C).
3. Two tubes (the microbial culture and nutrient broth containing thermometer) are placed into a thermostatically controlled water bath.
4. As soon as the NB temperature reaches to desired level, the time is recorded. Its temperature should be carefully monitored to ensure that it does not experience a significant change in temperature.
5. After holding time has elapsed, remove 1 mL of sample with pipette.

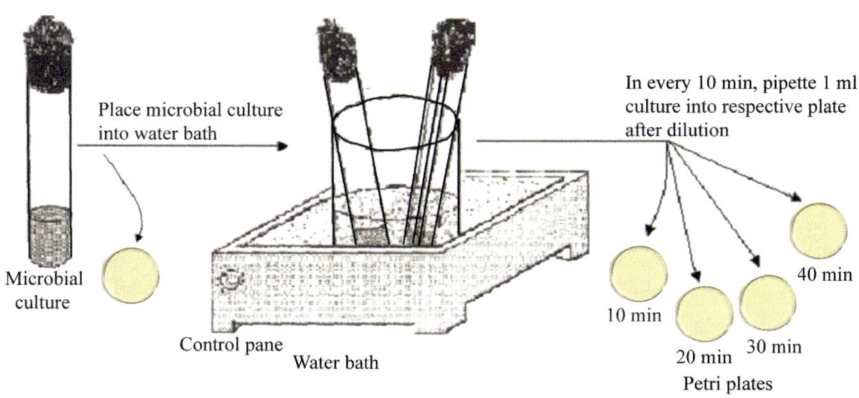

Fig. 20.2 Heating of medium at the specified temperature.

6. Dilute nonheated (control culture)/heated samples in the sterile 9 mL diluent (0.1% peptone water).
7. Remove 1 mL sample from nondiluted/diluted sample, inoculate into own labeled NA plate.
8. Repeat this operation for all intervals.
9. Incubate inverted plates at 35°C for 24 h in the incubator.
10. After incubation, examine plates and record the microbial growth results in your laboratory notebook.

Record your results from plates on the laboratory report with respect to time and temperature: No growth = none and growth = +. Determine TDT and TDP for each tested bacterium.

LABORATORY REPORT

Date:

Name:

Lab Title: Temperature Effect on Microorganisms

1. Record experimental results of microbial growth results from agar slant cultures.

	Temperature				
Microorganism	**5°C**	**25°C**	**35°C**	**45°C**	**50°C**

2. Record experimental results of microbial growth results from broth cultures.

	Temperature				
Microorganism	**5°C**	**25°C**	**35°C**	**45°C**	**50°C**

3. Record experimental results of temperature lethal effect on microorganisms.

	Temperature											
Microorganism	**5°C**				**25°C**				**35°C**			

	Temperature							
Microorganism	**45°C**				**50°C**			

4. **Based on the above data, determine the thermal death time and thermal death point for each microorganism**

Microorganism	Thermal death point	Thermal death time				
		5°C	25°C	35°C	45°C	50°C

5. **Explain the effects of temperatures on microorganisms below the minimum and above the maximum growth temperatures.**

6. **If a microorganism able to grow at 20°C, explain how you can determine experimentally whether the microorganism is a psychrophile, psychrotroph or a mesophile.**

7. **What is the optimum growth temperature for most human pathogens? Why?**

8. **Would you expect an endospore-former to be thermoduric? Why or why not?**

PRACTICE 21

Osmotic pressure and pH effects on microorganisms

21.1 Theoretical information

The growth of microorganisms is affected by the amount of water and hydrogen ions present in their environment. The amount of water entering or leaving the cell depends on the amount of water present in the environment. The transition of water molecules from a semipermeable membrane from a high amount of water content (low solute concentration) to a low amount of water content (high solute concentration) is called osmosis. Osmosis is a physical process that takes place with the effect of concentration through the semipermeable membrane without the energy use. Cell membranes are permeable to water.

There are three osmotic solutions depending on water content of environment: isotonic, hypotonic, and hypertonic. The isotonic solution contains the same concentration of dissolved matter as the amount of dissolved substance contained in the cell's cytoplasm. When a cell is placed in an isotonic solution, water molecules migrate into and out of the cells in equal amounts. Since cytoplasmic membrane permeable to water molecules. When microbial cell is placed into a hypotonic solution (low solute content in the environment), higher amount of water in the environment creates osmotic pressure on the microbial cell and the water molecules diffuse into the cell by osmosis. The water expands the cells and exerts pressure on the cell wall. This pressure is called turgor pressure. It can cause the cell to swell and possibly burst because the membrane can't withstand the high pressure inside the cell. When bacteria are placed in a hypertonic solution (high solute content in the environment), water molecules diffuse out the cells from cytoplasm by osmosis. The cells will shrink and dehydrate because of the loss of water from cytoplasm, plasma membrane is pulled away from the cell wall and probably cells die or inhibited. This event is called plasmolyzes. Some of the inactivated cells may return to normal form, when the cell is placed in an isotonic solution.

https://doi.org/10.1016/B978-0-323-91017-0.00009-3

When salt is added into microbial environment, salt in the environment creates hypertonic condition for most microorganisms, which undergo plasmolysis. It prevents microbial growth. Brine has been used to preserve a wide variety of foods (such as fish, cheese, meat, and pickles). Some microorganisms are halophiles; they require minimum concentration of salt. Obligate halophiles require a minimum of 13% NaCl in growth environment. Osmophiles, on the other hand, can grow at high concentrations of an organic solute, such as sugar. Some fungi can grow in honey or other concentrated sugar, and are known as osmophiles. Some microorganisms (such as molds and yeasts) growing in low-water environments are called xerophilic.

Bacteria are sensitive to changes in the hydrogen ion concentration of media. The concentration of hydrogen ions, which is designated by the term pH, limits the activity of enzymes. As the case of temperature, there is a minimum, optimum, and maximum growth pH for microorganisms (Fig. 21.1). The microorganisms show best growth rate at the optimum pH. The growth rate of microorganisms can be measured in terms of colony forming unit (cfu) per mL. The pH above which a microorganism fails to grow is known as the maximum pH, the pH below which a microorganism fails to grow is known as the minimum pH. The pH value of a given solution is the minus logarithm of the hydrogen ion concentration. A convenient method used to express hydrogen ion concentrations utilizes the well-known pH scale. The acidic range of the scale value extends from 0 to 6.9, while the alkaline or basic ranges from 7.1 to 14. Most microorganisms grow best at pH value around 6.5 or neutrality (7.0). Few microorganisms grow below 4.5. Bacteria tend to be more sensitive than fungi to pH changes in their environments. The susceptibility of microorganisms

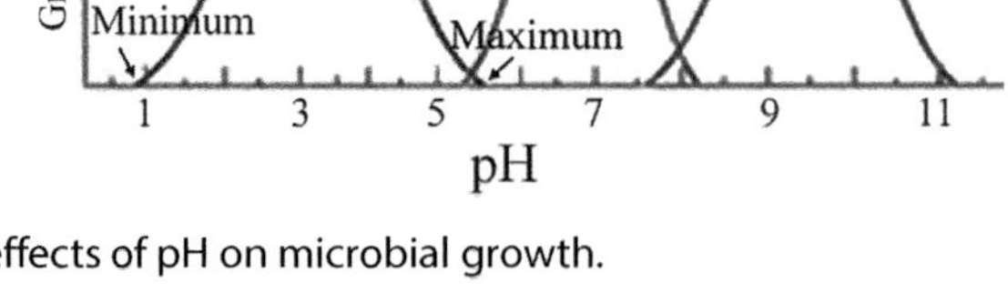

Fig. 21.1 The effects of pH on microbial growth.

can provide keeping quality for certain foods. It is quite common for fruits to undergo molds or yeasts spoilage because of the tolerance to pH values below 4.5. If the inorganic substance composition of medium, incubation temperature, oxygen amount and osmotic pressure of the environment changes, the hydrogen ion requirements of microorganisms can also change. In this practice, the effects of osmotic pressure and pH on the growth of microorganisms will be evaluated.

21.2 Effects of osmotic pressure and pH on microorganisms

21.2.1 Effect of osmotic pressure on microbial growth

Materials

1. 24 h nutrient broth cultures of *Escherichia coli*, *Staphylococcus aureus*, and Halobacterium halobium.
2. Trypticase soy agar (TSA) plates with 1, 5, 10, and 25% glucose, and TSA plates with 1, 5, 10, and 20% NaCl.

Procedure

With each culture provided, perform the steps described in the following procedure.

1. Label the plates with glucose and NaCl concentrations, and your initials. Invert the plates and divide each into three sections with straight lines and label each sector with the given bacterial culture.
2. Streak plate a loopful of the culture onto the marked section of the plate. Continue until all the remaining cultures have been streaked.
3. Incubate inverted plates at optimum growth temperature from 24 to 48 h.
4. After incubation, look at for the growth of bacteria on plates. Record the results on your laboratory notebook. Use the following system for indicating the extent of growth: No growth = none, poor growth = +, good growth = ++ or excellent growth = +++.

21.2.2 Effects of pH on microbial growth

Materials

1. 48 h nutrient broth bacterial cultures of *E. coli, Bacillus subtilis*, and *Alcaligenes faecalis*, and 5 days potato dextrose broth yeast culture of *Saccharomyces cerevisiae*.
2. Glucose phenol red broth (GPRB, 5 mL) in the tube and potato dextrose broth (PDB, 5 mL) in the tube with pH of 3, 5, 7, 8, and 9. Loop.

Procedure

With each culture provided, perform the steps described in the following procedure.

1. Note the variations in color among the GPRB with different pH value: pH indicator phenol red is orange at pH 7.0, yellow at pH below 7.0, and red at pH above 7.0.
2. Label each tube with pH value, color, microorganism, your initial names, and type of media.
3. Inoculate GPRB tubes with one loopful of each bacterial culture and PDB tube with yeast culture.
4. Incubate all bacteria inoculated tubes at 35°C for 24 h and yeast inoculated tubes at 28°C from 1 to 3 days.
5. After incubation, examine each tube for the growth of microorganisms. Record the results in your laboratory notebook. Use the following system for indicating the extent of growth: No growth = none, poor growth = +, good growth = ++ or excellent growth = +++. Note also any color changes in the medium.

LABORATORY REPORT

Date:

Name:

Lab Title: Effect of Osmotic Pressure and pH on Microbial Growth

1. Record results from osmotic pressure effects.

	Glucose concentration %				Salt concentration %				
Microorganism	1	5	10	25	1	5	10	15	20

2. Record results from pH effects.

Microorganism	3	5	7	8	9

3. How does a solid concentration affect the growth and survival of microorganisms?

4. How do acidic and alkaline environments affect the growth and survival of microorganisms?

5. **Define followings and give least one example from experimental results and to food providing from these environments:**

   Halophilic:

   Osmophilic:

   Xerophilic:

6. **Answer the following questions with the information concerning sugar and salt containing agar medium:**

   (a) For which microorganism is the medium selective? Indicate selective concentration for each microorganism.

   (b) Which ingredient makes medium selective for microorganisms?

7. **Which type of solution can be used in food preservation? Why?**

PRACTICE 22

Ultraviolet light effects on microorganisms

22.1 Theoretical information

Some radiations are gamma-ray, X-ray, and UV light. Most bacteria are harmed by UV light. UV light is absorbed by many intracellular compounds in bacterial cells, but DNA suffers the most damage. The greatest bactericidal action occurs at the UV wavelengths around 260 nm, which are the wavelengths absorbed most strongly by DNA. After DNA is exposed to UV light, pyrimidine dimers form. DNA replication can be inhibited or altered. This can cause death or mutation which in turn causes changes or losses of associated properties of microorganisms. Examples of such mutations include changes in colony size, loss of capsule, loss of pigmentation, and a significant reduction in viability cells (lethal effect). Special lamps that emit high intensity of microbicidal UV wavelengths are used as the source of UV light to kill microorganisms. It can only be used in very limited application areas since it has a very weak transition capacity through materials. Only microorganisms on the surface of an object are likely to be killed by UV radiation. Even a thin layer of glass or water can stop effect of UV light. This form of radiation is commonly used to reduce the number of microorganisms in air, on surface area of equipments and processing rooms, on the surface area of natural drinking water and in aseptic rooms where sterile products are being dispensed into sterile vials. UV wavelengths application for these purposes are in the range of 250–265 nm. Wavelengths of light may be expressed in nanometers (nm) or angstrom (Å = 0.1 nm). UV light includes electromagnetic radiations that fall in the wavelength band between 4 and 400 nm. The visible range is approximately between 400 and 780 nm.

Antiseptic effects of UV light on surface areas and in air occur only in a certain UV ray spectrum. As stated in Fig. 22.1, the most effective wavelength of UV is 265 nm. Low pressure mercury vapor lamps with high efficiency are used for ultraviolet application. In this practice, microorganisms will be exposed to UV radiation over an agar medium for various length of times to determine the minimum exposure time to achieve killing of microorganisms.

Laboratory Practices in Microbiology
https://doi.org/10.1016/B978-0-323-91017-0.00010-X

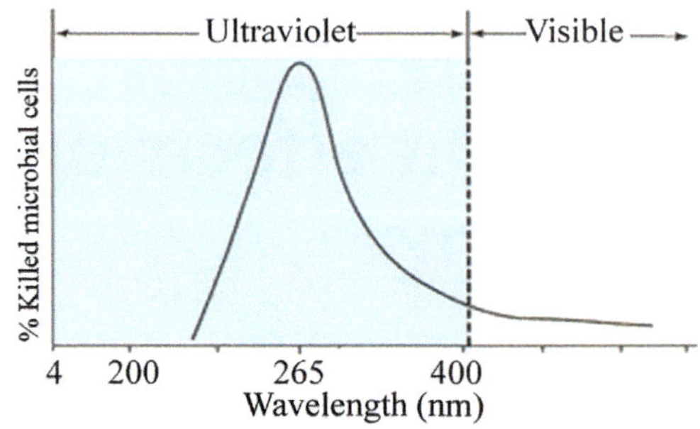

Fig. 22.1 Lethal effect of ultraviolet light.

22.2 Ultraviolet light application

Materials

Microbial cultures: 48h nutrient broth cultures of *Escherichia coli*, *Staphylococcus aureus*, and *Bacillus subtilis*.

Media and equipments. Nutrient agar plates, UV lamp, sterile pipettes (1 ml) in box, alcohol in 50mL beaker and glass spreader.

Procedure

With each culture provided, perform the steps described in the following technique.

1. Label bottom of each plate with a given bacterial culture and your initials.
2. Turn on UV lamp. Place the lamp about 9cm space from the surface of Petri plate (Fig. 22.2).

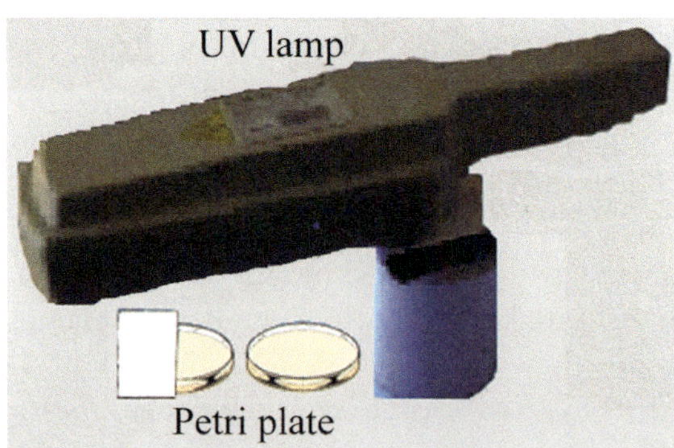

Fig. 22.2 Expose plates to UV light.

3. Draw a line dividing the bottom of each plate into two sections. Label each half with one of the following exposure times: None, 5, 10, 15, 20, and 30 min.
4. Aseptically transfer 1 mL of broth culture onto the surface area of each plate using sterile pipette.
5. Blood the glass spreader into alcohol in the beaker, flame the glass spreader and allow it cool near flame.
6. Spread the inoculum on the agar surface using the glass spreader.
7. Allow to dry the surface area of agar in the Petri plate.
8. Remove the lid of plate and expose the labeled half part of plate to UV for specified time period. The plate including "none" as a label, remove the lid of plate and cover this half part of the plate with a card that is not to be irradiated. *Note*: Do not look directly at the lamp, as severe eye damage can result.
9. After the plates have been exposed to UV light, replace the lids of Petri plates.
10. Incubate inverted Petri plate at optimum growth temperature of microorganisms for 48 h.
11. After incubation, examine each tube for the growth of microorganisms. Record the results in your laboratory notebook. Use the following system for indicating the extent of growth: No growth = none, poor growth = +, good growth = ++ or excellent growth = +++. Note also any color changes in the medium.
12. Record the results in your laboratory notebook. Determine the minimum amount of exposure required to 100% kill.

LABORATORY REPORT

Date:

Name:

Lab Title: Ultraviolet Light Lethal Effect on Microorganisms

1. **After incubation, observe the colony formations that will occur in Petri plate and indicate extent of growth.**

	UV light exposure time (min)					
Microorganism	**0**	**5**	**10**	**15**	**20**	**30**

2. **Does UV have differential effect on Gram-negative and Gram-positive microorganisms based on your data? Why?**

3. **Why do endospores not affected as well as the vegetative cells exposed to UV rays have a lethal effect?**

4. **What limited protection does the microorganism have against the harmful effects of UV radiation?**

5. **Which bacteria would show the best survival following exposure to UV radiation? Why?**

6. **What wavelength of ultraviolet is most germicidal for microorganisms?**

7. **Explain importance of UV light for food and water industry.**

8. **What is the difference in the mode of action of UV rays and gamma rays?**

PRACTICE 23

Oxygen effects on microorganisms

23.1 Theoretical information

Microorganisms can be listed to five groups according to the oxygen (O_2) requirements: (i) aerobic microorganisms grow in the presence of O_2, (ii) facultative anaerobic microorganisms grow either in the presence or absence of O_2, (iii) microaerophilic microorganisms require limited amounts of free molecular O_2 (2%–10%), (iv) anaerobic microorganisms grow in the absence of O_2, and (v) aerotolerant anaerobic (fermentative) microorganisms grow in the presence of O_2 but can't use O_2 in metabolic system and obtain energy only by fermentation. Fluid thioglycolate semi-soft medium can be used to determine O_2 requirements of microorganisms. In the preparation of medium, sodium thioglycolate is added as a reducing agent and methylene blue as an oxidation-reduction indicator. Sodium thioglycolate possesses sulfhydryl groups (SH), which tie up O_2 molecule, and prevent the formation and accumulation of H_2O_2 in the medium. Oxygen tension is always higher near the surface area of the medium. Medium will be colored at the top because of presence of O_2, and colorless in the middle and bottom, because of absence of free O_2. The medium contains lower amount of agar (0.5%) that restrict the microorganisms to small area and provides anaerobiosis through medium (in the middle and at the bottom) in tube. Anaerobic agar which is another medium contains thioglycolate as a reducing agent and resazurin as an oxidation/reduction indicator. For strict anaerobic microorganisms, medium must be incubated in an oxygen-free anaerobic incubator or anaerobic jar. In anaerobic incubator, O_2 is replaced with CO_2 in the incubator space to provide anaerobic condition.

Gas pack is placed into the anaerobic jar to provide anaerobic condition. The reaction of the palladium in the Gas pack is catalyzed at room temperature by adding water into the pack. This rection produces hydrogen. The hydrogen in the jar creates water by combining with O_2 and this reaction provides anaerobic conditions. Anaerobic conditions that will occur in the jar are understood by the decolorization of methylene blue in the jar. If methylene blue does not decolorize within 2 h, the jar may not be properly

Laboratory Practices in Microbiology
https://doi.org/10.1016/B978-0-323-91017-0.00015-9

closed, no chemical reaction has occurred or it may be result from other effects. In these cases, new Gas pack must be placed into jar. In this practice, the effects of O_2 on the growth of microorganisms will be indicated.

23.2 Determination of oxygen effects on microorganisms

With each culture provided, perform the following techniques in indication of oxygen effects on microorganisms.

Materials

Microbial cultures: 24h nutrient broth cultures of *Escherichia coli*, *Staphylococcus aureus*, *Brochothrix thermosphacta*, *Clostridium perfringens*, *Lactobacillus bulgaricus*, and *Yersinia enterocolitica*.

Media. Fluid thioglycolate medium (FTM), tryptone glucose yeast extract agar (TGYEA) plates, and Brewer's anaerobic agar plate.

Equipments. Water bath at boiling, loop, anaerobic jar, Gas pack, scissor, sterile water in Erlenmeyer flask, and sterile pipette (10 mL) in box.

23.2.1 Growth of microorganisms in FTM

1. The tubes are labeled with the name microorganism, your initials, and the name of medium.
2. To prevent O_2 entering FTM in the tube, the tube should be kept steadily, without shaking. If the top of the FTM in the tube is pink at 30%, it will be an indicator of the presence of O_2, so the medium must be boiled in the tube for a few minutes to remove O_2 from the medium.
3. Hold the tubes gently and without shaking to prevent unwanted O_2 entering into the FTM.
4. Inoculate each of FTM in the tubes with one loopful of microbial culture by inserting into medium.

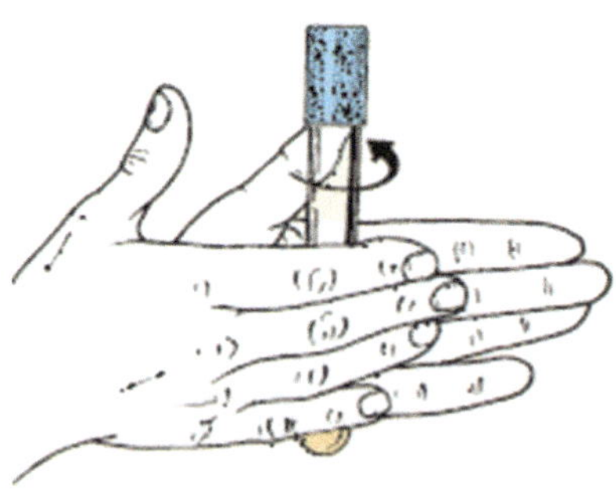

Fig. 23.1 Microorganism and medium are rolled in the tube with your hands to mix.

5. Gently roll each tube in the palm as shown in Fig. 23.1 to ensure good distribution of microorganisms in the medium.
6. Incubate FTM tubes at 35°C for 24–48 h in an incubator.
7. After incubation, be careful not to shake the FTM tubes when carrying the receptacle and tubes to your table. If you shake, microorganism growth in FTM can easily be proliferated.
8. Evaluate FTM and record results in your laboratory report. Indicate the microorganism as being *aerobes*, *microaerobes*, *facultative anaerobes*, *anaerobes*, or *aerotolerant anaerobes*.

Aerobic microorganisms require O_2, FTM contains O_2 near the surface, therefore show growth near the surface of FTM (Fig. 23.2a). Aerobic microorganisms cannot grow without an abundant supply of O_2; therefore, they cannot grow in the rest of FTM.

Microaerobic microorganisms require a minimum level of oxygen (5%–10%) for growth. FTM contains just right concentration of O_2 near below the surface area, therefore they grow below the surface area of FTM (Fig. 23.2b).

Facultative anaerobic microorganisms (or facultative aerobes) can grow in the presence or absence of free O_2, but generally show heavy grow in the presence of O_2. They grow throughout the medium but better below the surface area of FTM (Fig. 23.2c).

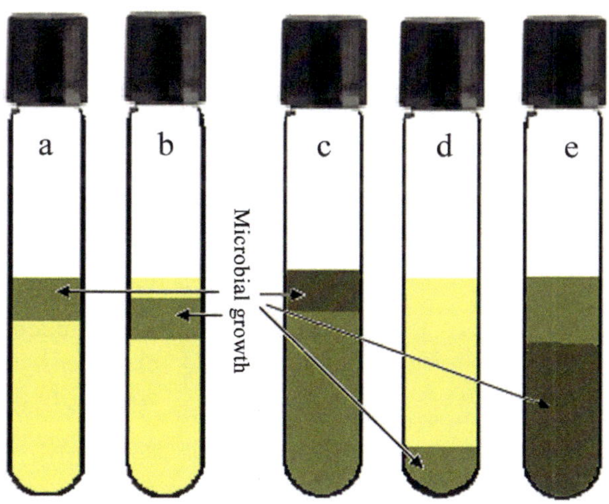

Fig. 23.2 Growth patterns for different types of bacteria in FTM according to O_2 requirement. Growth of aerobes (a), microaerobes (b), facultative anaerobes (c), anaerobes (d) and aerotolerant anaerobes (e).

Anaerobic microorganisms grow in the absence of O_2. They show heavy growth at the bottom of the tube and not growth remaining part of the FTM (Fig. 23.2d).

Aerotolerant anaerobic (fermentative) microorganisms can grow in in the absence of O_2, but presence of O_2 cannot affect the growth of these microorganisms, do not use O_2 to grow. Therefore, they can grow throughout the medium, but better in the absence of O_2 (Fig. 23.2e).

23.2.2 Anaerobic incubation in an anaerobic jar

1. The TGYEA plates are labeled with the microorganisms assigned, your initials, and the name of medium.
2. Using a water-proof marking pen divides the Petri plates into six sections and label with the name of microorganism.
3. Streak microorganisms on the Petri plate onto labeled section using a loopful liquid culture.
4. Place the inverted Petri plate into the anaerobic incubator or the anaerobic jar (Fig. 23.3).

   (a) Incubation in the anaerobic jar

   To activate and seal the anaerobic jar as follows:

   i. Place a gas pack into the jar.

   ii. Cut the corner of each gas pack with the scissor. Place packages with cut corners upright in the jar.

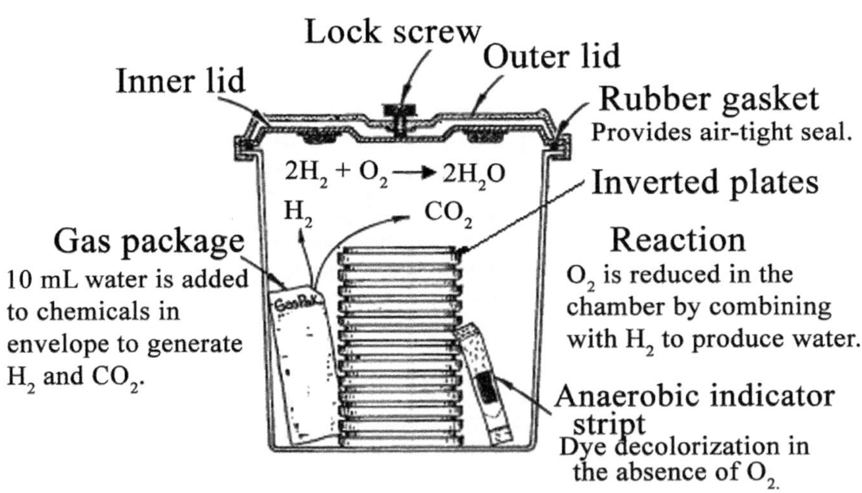

Fig. 23.3 Incubation in the anaerobic jar.

iii. Add 10 mL of pure water into the package with the pipette from the opened corner of the package. When adding water, avoid forcing the package to the envelope by pipette.
iv. Place the lid of the jar. Do not use grease or any other sealant on the edge of the jar, the O-ring on the lid provides an effective seal when pressed on a clean surface.
v. Close the lid of the jar and place the jar into an incubator at 35°C for 24–48 h. After 2 or 3 h, check if the indicator strip inside the jar has lost its blue color. If no color change has occurred, replace the palladium, and repeat the entire process.
vi. Within 30 min after closing jar, droplets of moisture will begin to condense on the inside of the jar, indicating that hydrogen has been generated and chemically combined with the atmospheric oxygen, aided by the catalyst, to form water.
vii. After incubation of jar in the incubator, open the gas valve of the jar and to avoid pressure in the jar, open the jar and remove the gas pack. If the jar is kept tightly in place because of the vacuum formed inside the jar, a vacuum can be taken by sliding the lid aside.

(b) Incubation in anaerobic incubator

i. Place inverted plates into anaerobic incubator. Close the door of incubator and adjust the anaerobic incubator to CO_2 level.
ii. Incubate the plates at 35°C for 24 h under anaerobic conditions.

5. After incubation, indicate growth patterns on Petri plates for types of microorganisms depending on O_2 requirement. Examine your plates carefully for the growth of microorganisms on the Petri plate. Write a brief description of the amount and type of growth for each microorganism in your laboratory notebook.

23.2.3 Microaerophilic incubation

A candle jar can be used for the cultivation of microaerobes (such as *Streptococcus pyogens*, *Brochothris thermosphacta* and *Campylobacter jejuni*).

1. The TGYEA plates are labeled with the microorganisms assigned, your initials, and the name of medium.
2. Using a water-proof marking pen divides the Petri plates into six sections and label with the name of microorganism.
3. Streak microorganisms on the Petri plate onto labeled section using a loopful liquid culture.
4. Inoculate TGYEA plates, place inverted plates into the jar.

5. Place a candle into the jar, fight the candle and tightly close candle jar. Light the candle and seal the jar tightly. A 5%–10% CO_2 atmosphere will be created in the jar. Burning of candle reduces amount of O_2 to 5%–10% and this provides microaerobic environments in the jar.
6. After incubation, examine your plates carefully for the growth of microorganisms on the Petri plate. Write a brief description of the amount and indicate microaerobic microorganisms in your laboratory notebook.

LABORATORY REPORT

Date:

Name:

Lab Title: Oxygen Effects on Microorganisms

1. **Draw the growth seen in each of five FTM culture according to your results.**

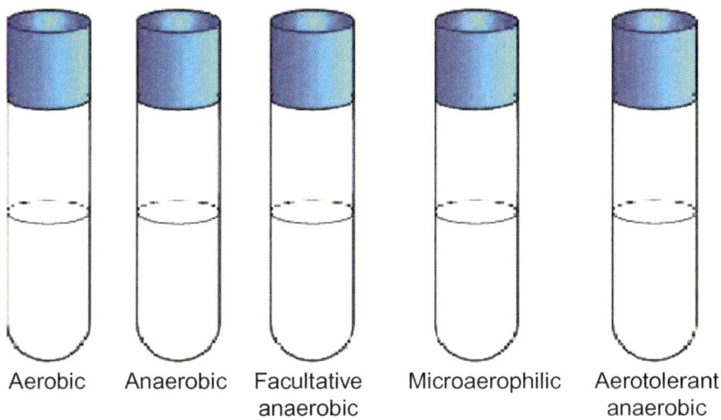

2. **Why can media (such as thioglycolate) be used for cultivation of anaerobes?**

3. **Indicate type of growth in the anaerobic incubator (or anaerobic jar) which you used in the practice. Explain reasons of growth and without growth.**

4. Indicate type of growth in the candle jar which you used in the practice. Explain reasons of growth and without growth.

5. What is the role of oxygen in cellular respiration?

6. What type of metabolism occurs in the absence of oxygen?

7. Explain importance of microbial group in foodborne disease, food spoilage and food production.

SECTION V

Fungi and bacteriophages

Fungi vary widely in size and shape from unicellular (microscopic) to multicellular (macroscopic) structures. Individual cell of fungi ranges from 10 to 100 μm. Yeasts are single cellular microorganisms. Molds are multicellular and form hyphae. Hyphae form filamentous structure. Bacteriophages, known as bacterial viruses, replicate within bacterial cells. Generally, bacteriophages show a unique relationship to certain species of bacteria. As with all viruses, bacteriophages consist of a genetic material surrounded by a protein capsid. The genetic material, nucleic acid, can be DNA or RNA, and the genetic material can be double-stranded or single-stranded. A bacteriophage (phage) can infect only one type or several types of bacteria. In general, there is specificity between bacteriophages and bacteria. The structure of bacteriophages can be displayed in detail by an electron microscope. Bacteriophage has three basic structures: a tailed icosahedral (20-sided) head, a non-tailed icosahedral head, and a filamentous part. Through representative practices, this section will examine the morphological characteristics of molds and yeasts as well as introduce the isolation, enumeration, and uses of bacterial viruses.

The practices presented in this section are as follows:

Practice 24. Morphological Characteristics of Yeasts and Molds

Practice 25. Isolation, Enumeration, and Uses of Bacteriophages

PRACTICE 24

Morphological characteristics of yeasts and molds

24.1 Theoretical information

Fungi (singular: fungus) are eukaryotic microorganisms that include yeasts and molds (British English: molds), as well as the mushrooms. These microorganisms are classified in the Fungi kingdom. The fungal cells have cell walls as well as bacteria but structure of fungi walls contains chitin, unlike the cell walls of plants, which contain cellulose. The discipline of biological science of fungi is known as mycology. A mass of hyphae is called a mycelium. When hyphae have cross-walls between cells, they are called septated hyphae (singular septum). When hyphae do not have cross-walls between cells, they are called nonseptated hyphae.

The fungi include yeasts and molds. Fungi media (mycological media) differ from bacterial media with different requirements of fungi. Most fungi can grow at lower pH but most of the bacteria cannot grow. Fungi grow on media containing inorganic salts in the presence of carbohydrate as an energy source. Some fungi require B-group vitamins or other growth factors in the growth medium and the medium may be provided by adding 0.1% yeast extract. Almost all molds are obligate aerobes. There is no special staining or biochemical tests to identify the mold genus, their structures are identified microscopically. The structural differences that are most often helpful in the identification of filamentous fungi (molds) are the different types of spores, spore arrangement and hyphae. Molds differ from yeasts and most bacteria in the following ways:

1. They have different colony morphology.
2. They have filamentous structure.
3. They are microscopic with mycelium.
4. They are multicellular.
5. They have specialized structures with specialized function.
6. Their asexual reproduction forms from fragmentation of special structures.

In nature, the principle role of fungi is their power of degradation that reduces organic material to inorganics. The individual filaments of molds

https://doi.org/10.1016/B978-0-323-91017-0.00022-6

are called hyphae (hypha, singular). If the hyphae of mold have cross-walls between cells, it is called as septated hypha. If cross walls are not present between cells, the hyphae are called nonseptated. A mass of hyphae is macroscopic and is called mycelium. The placement of various types of spores on vegetative hyphae is particularly important. When someone tries to prepare a wet surface by transferring hyphae from the colony to the slide, the destruction of the spores pattern and structure should be prevented.

Gram stain of yeast culture can allow the appearance of asexual buds. Yeasts colonies, unlike the molds, look more like bacteria on media. The yeast will differentiate with physiological activities, for example, *Saccharomyces cerevisiae* converts simple sugars to ethyl alcohol and carbon dioxide under anaerobic condition. Most of the yeasts of industrial importance are members of the genus *Saccharomyces*. Yeast cells are round, short-oval, single, pairs or short chains or groups. Yeasts can reproduce by a process called budding. A bud or blastospore forms as an out pouching of the parent cell. In some yeast, buds may remain attached to each other to form a pseudohyphae and they are called dimorphic. Yeasts cannot have hyphae. During the proliferation of yeast, a small bud can be easily distinguished from the main cell. Some yeast can also produce by binary division.

Kingdom fungi (Myceteae) has six main classes: Zygomycota, ascomycota, basidiomycota, deuteromycota, oomycota, and myxomycota.

- **Zygomycota.** They have nonseptate hyphae and produce zygospores and sporangiospores. *Rhizopus*, *Mucor*, and *Syncephalastrum* are examples of genus for this class. Zygomycota spores can be formed sexually and asexually.
- **Ascomycota.** All fungi in this class produce ascospores and are called "sac fungi." All of them have septated hyphae. They produce single ascus. *Eupenicillium*, *Penicillium chrysogenum*, *Candida albicans*, *Aspergillus flavus*, *Ophiostoma ulmi* and *Cryphonectria parasitica* present in this class.
- **Basidiomycota.** They produce basidiomycetes such as puffballs, mushrooms, smuts, rust, and shelf fungi. Basidiomycota are filamentous fungi composed of septated hyphae (except for basidiomycota-yeast) and reproduce sexually via the formation of specialized club-shaped end cells called basidia that normally bear external meiospores (usually four). These specialized spores are called basidiospores. However, some Basidiomycota are obligate asexual reproducers. Basidiomycota can also reproduce asexually. *Agaricus bisporus* and *Lentinus edodes* are present in this class.
- **Deuteromycota.** The asexual and sexual stages of fungi are present. Genera of this group are called as the fungi imperfecti. Fungi imperfecti

does not fit into the commonly established taxonomic classifications of fungi that are based on morphological characteristics of sexual structures because their sexual form of reproduction has never been observed. These fungi produce their spores asexually, in the process called sporogenesis. Among the economical important genera of the imperfect fungi are *Penicillium* and *Aspergillus.*

- **Oomycota.** The name "oomycota" means "egg fungus". They are plant pathogens and can cause economical loss. Many of them have motile spore which can swim by lateral flagella. They produce asexual spores (such as oospores). The oomycetes rarely have septa and sporangia. Some genera are unicellular fungi, and other genera are filamentous and branching. In sexual reproduction, the male nuclei (sperm) are directly injected into the oogonium. *Pustula*, *Wilsoniana*, *Phytophthora*, and *Plasmopara viticola* present in this class.
- **Myxomycota.** They include slime molds. They produce plasmodia, hairy material, and fruiting structure of this class. The myxomycota differs from true fungi because their vegetative body consists of only protoplast bounded by membrane and devoid of cell wall. This class has *Arcyria cinerea*, *Trichia varia*, *Hemitrichia clavata*, and *Hemitrichia vesparium*.

Molds are differentiated and characterized based on the type of spore formation. Molds produce sexual and asexual spores.

(i) Asexual spores. The following asexual spore types are shown in Fig. 24.1:

- **Phialospores.** Vase-shaped cells produce conidia spores and called phialides. *Penicillium* and *Gliocadium* produce this type of spores.
- **Blastospores.** They are produced within the phylum with the formation of buds from cells, as in *Cladosporium,* which typically has lemon-shaped spores. An example of unicellular fungus that forms blastospore is *Candida albicans*.
- **Arthrospores.** Conidia forms from vegetative cells and pass into resting state. Arthrospores are variable in wide and length, and cylindrical. Spores are produced by fragmentation of vegetative hyphae. Examples of the mold genera producing arthrospores are *Mucor* and *Geotrichium*.
- **Chlamydospores.** These kinds of spores are found in large, thick-walled, round or irregular structures that form inside or at the end of the hyphae. They are common to most fungi, usually occur in old cultures. They are usually dark-colored, spherical and have a smooth surface. Examples of chlamydospore forming fungi are *Trichosporon* and *Cryptococcus*.

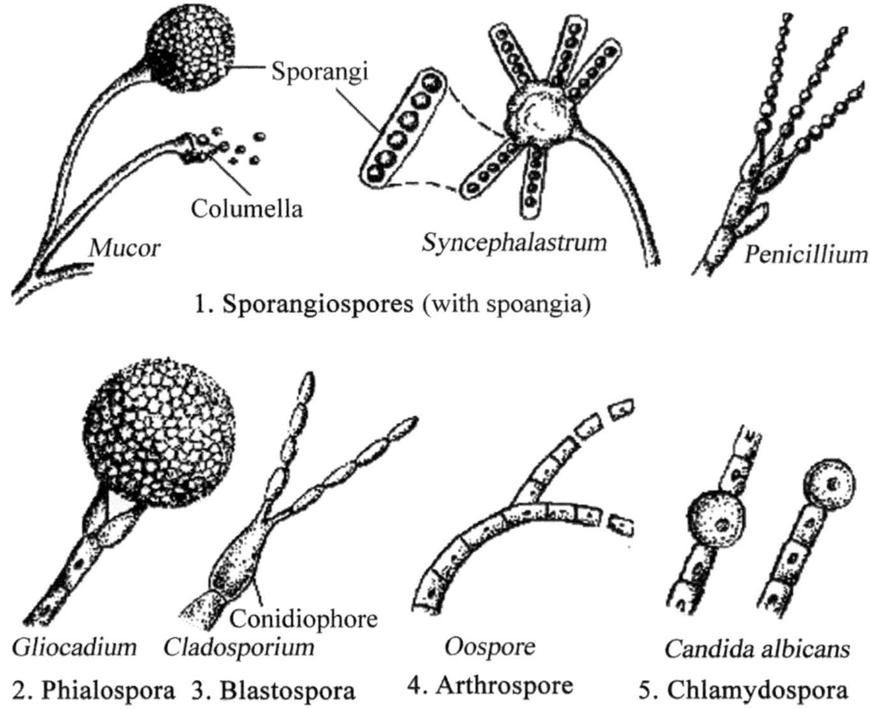

Fig. 24.1 Asexual spore types produced by fungi. *(Modified with permission from McGraw-Hill; H.J. Benson, Microbiological applications: laboratory manual in general microbiology, 2001, Fig. 10.2, p. 49.)*

- **Sporangiospores.** These spores are produced in a sporangium. During the activation of the spores, they swell by taking water and form a spherical sporangium. Its nuclei are divided by mitosis many times until the sporangium fills the nucleus. Each core is then surrounded by a spore wall and filled with spores. Sporangia can produce spores by mitosis. *Rhizopus oligosporus*, *Rhizopus domesticus*, and *Absidia spinosa* produce sporangia spores.

(ii) Sexual spores. Sexual spores of fungi are oospores, zygospores, ascospores, and basidiospores. Fig. 24.2 illustrates the types of sexual spores:

Oospores. They are produced when male gametes (reproductive nuclei) enter a large cell (oogonium) and fertilize the eggs.The oogoniums are usually penetrated by one or more hyphae that give rise to the male nuclei. *Phytophthora infestans*, *Solanum demissum*, and *Hyaloperonospora parasitica* are examples of oospore forming molds.

Zygospores. Zygospores are not produced in any kind of enclosing structure of hyphae. Direct fusion of two hyphae from neighboring filaments produces zygospores. In general, zygospores are large, spherical, often

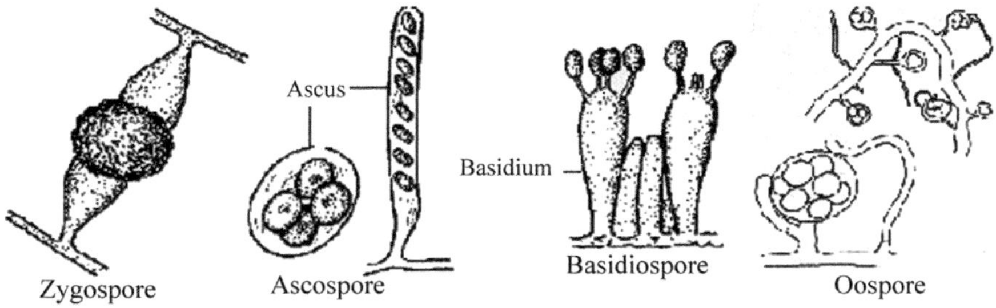

Fig. 24.2 Types of sexual types of spores produced by fungi. *(Modified with permission from McGraw-Hill; H.J. Benson, Microbiological applications: laboratory manual in general microbiology, 2001, Fig. 10.3, p. 50.)*

dark brown or black colored, rough-walled spores with two connecting hyphae. *Mortierella capitata*, *Orphella catalaunica*, *O. coronata*, and *O. helicospora* are examples of zygospore forming mold.

Ascospores. An ascospore is a spore in an ascus. Ascospores are created in cylindrical cells called asci. These are mostly in groups of four or more spores. They are usually produced in cells that have an asci-type structure and therefore do not form hyphae. Ascospore fertilization takes place in the early sporulation period. *Morchella elata* and *S. cerevisiae* can be given as examples of mold-forming ascospore.

Basidiospores. The structure that creates sexual spores from a structure is called Basidium. Four basidiospores are formed in the extension of each basidium. A typical mushroom produces billions of basidiospores. *Uromyces appendiculatus* and *Agaricus bisporus* are examples of basidospore forming molds. Some of the yeast can also present in this group without spore formation, such as *Schizophyllum commune*.

24.2 Morphological characteristics fungi

24.2.1 Morphological characteristics of yeasts

Materials

Microbial culture: 24 h potato dextrose broth culture of *S. cerevisiae*.

Media: Sabouraud dextrose agar (SDA) plates, Sabouraud dextrose broth (SDB).

Equipments: Glass-slides in bottle containing alcohol, coverslips, loop, microscope, and immersion oil.

(1) Morphology of yeast from liquid medium

1. Inoculate one loopful of culture into SDB and incubate at 25°C for 1–5 days in an incubator.

2. After incubation, transfer two or three loopful of the yeast broth culture on a slide with a drop of methylene blue stain and cover lightly with a cover slip to prepare a wet-mount.
3. Focus with the low-power objective; then use the high-power objective.
4. Examine the yeast cells carefully and look for the nucleus, granules, cell walls, buds, vacuole, and others. Oil-immersion objective will provide the greatest detail magnification. The yeast cells were analyzed for the nucleus and vacuole. The nucleus appears as the smaller body within cells. Draw some cell morphologies on your laboratory notebook.

(2) Microscopic and Cultural Characteristics of Yeast

1. Shake *S. cerevisiae* culture tube gently to disperse cells before use and streak plate onto SDA plate using loop.
2. Incubate inverted plates at 25°C for 2–5 days in the incubator.
3. After incubation, examine colony morphology and make drawings of several isolated colonies of yeast.
4. Record colony morphology and color.
5. Make a Gram stain from yeast colony and analyze using oil-immersion objective.
6. Make drawings of the microscopic appearance and look for budding cells. Draw cells on your laboratory notebook.

24.2.2 Mold morphological characteristics

Morphological characteristics of molds include color changes, colony appearance, hyphae, mycelium, spores and arrangement of hyphae. Morphological analysis can also predict septation of hyphae.

Materials

Microbial culture: 24h Sabouraud dextrose agar plate culture of *Aspergillus niger.*

Medium and others: Sabouraud dextrose agar (SDA) plate and alcohol in 50m beaker lactophenol cotton blue stain and sterile water in tubes.

Equipments: U-shaped glass rod, sterile filter paper (90mm in diameter) in Petri dish, glass slides in bottles containing alcohol, cover slips, microscope, forceps, loop, and sharp scalpel.

(1) Cultural characteristics

1. Inoculate *A. niger* culture by streak plating the surface area of medium using cotton swab.
2. Incubate inverted plates at 25°C for 3–5 days in the incubator.

3. After incubation, observe colony morphology:
   (a) *Macroscopic observations*: Examine mold cultures on SDA. SDA medium has low pH, which makes the medium selective for molds. A good Petri plate provides many colonies in different colors. The original "cottony" appearance of the colonies is an important feature. Also, looking at the Petri plates, it can be observed how the colonies differ in color. The definition of molds is made according to surface color, Petri plate color, hyphae structure, and spore properties.
   (b) *Microscopic observations*: Place the Petri plate on the microscope field and examine the edge of a colored colony with a low-power objective. Determine the hyphal structure and spores. The white colonies that may be encountered during the image should be ignored, because their spores are often not found and difficult to identify.
4. Indicate colony morphology and spore types.
5. Transfer a small amount colony from plate with a loop onto a flamed glass slide and add a drop of lactophenol blue dye to the edge of colony. Cover this preparation with the cover glass and examine it with a low-power objective and then with a high-power objective under a light microscope. Record hyphae morphology. Draw colony morphology.

(2) Preparation of a mold microculture

Prepare slide culture in the following manner and shown in Fig. 24.3.

1. Using disinfected forceps, place a sterile filter paper into sterile Petri dish bottom under aseptic conditions.
2. Disinfect a U-shaped glass rod by blotting into alcohol and flaming.
3. Place disinfected glass rod on the filter paper.
4. Add about 5–10 mL sterile water on filter paper to completely moisten filter paper. It prevents dehydration of agar medium during incubation.
5. Place microscope slide (disinfected with burning alcohol) on the U-shaped glass rod with pens.
6. Gently flame a scalpel containing alcohol to sterilize, and now cut a block of SDA (1 cm^2) from SDA plate with scalpel (or with flamed inoculating loop).
7. Inserting a disinfected scalpel into one side of agar and remove the agar block. Inoculate agar block at top and bottom surfaces with mold spores using loop from the mold colony on agar medium.

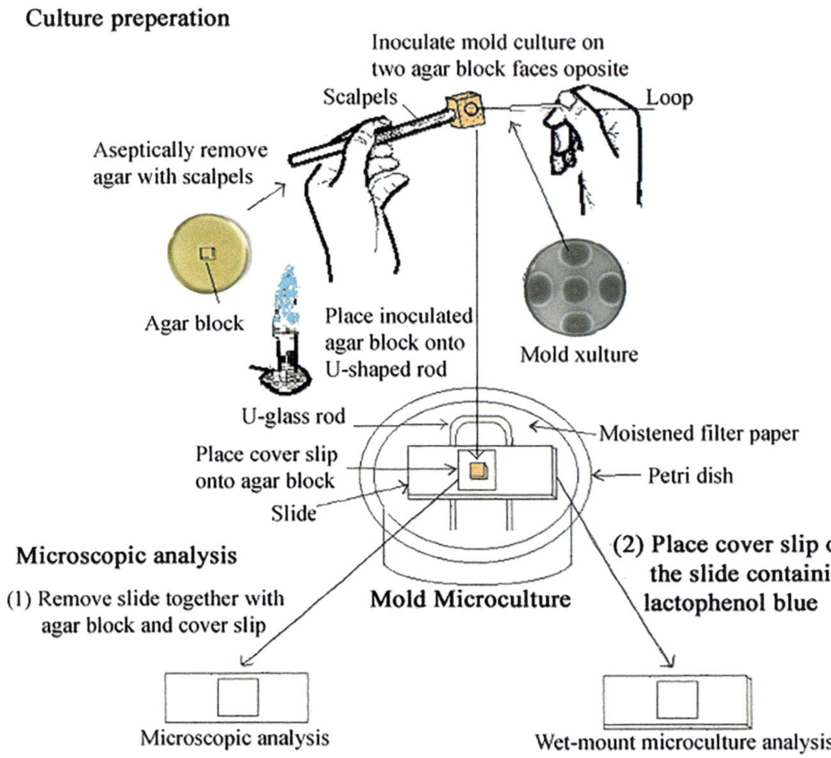

Fig. 24.3 Preparation of the mold microculture and microscopic analysis.

Be sure to flame the loop prior to picking up spores from mold culture on PDA.

8. Place the inoculated agar block, inoculated sides down and up, onto the U-glass rod in the Petri dish.
9. Sterilize a cover slip in the following manner while holding the cover slip with a slide forceps; dip the cover slip into a small beaker containing alcohol. Let most of the alcohol drain off and then quickly pass the cover slip through the flame of the Bunsen burner once. This will burn off the remaining alcohol and completes the disinfection process.
10. Aseptically, place the disinfected cover slip on top of the inoculated agar surface block.
11. Place the cover of the Petri dish and incubate at 25°C for 24–72h in the incubator.

12. After incubation, observe colony morphology:

(a) *Microscopic observation.* After 24, 48 or 72 h, if growth is adequate on agar medium, remove the slide from the Petri dish and place it on the low power objective (agar block, cover slip and microscope slide together). If growth has occurred, you will see hyphae and spores. If growth is inadequate and spores are not evident, allow the mold to grow another 24–48 h incubation before making the stained slides. After incubation, magnify slide again for hyphae and spore formations.

Using both the low- and high-power objectives, look at carefully the hyphae structure and spores. Make microscope drawings of the specialized structures (hyphae, mycelium, and spores on the complete structure of mold) in your laboratory notebook and label them.

(b) *Microculture observation*

Place 2 drops of lactophenol cotton blue on a clean microscope slide.

With forceps, remove the cover slip from the microculture agar block after microscopic analysis.

Place the cover slips with culture growing side clinging on the pool of lactophenol cotton blue on the slide. Forceps must be disinfected before and after use with blotting into alcohol and flaming.

Precaution: To avoid trapping large air bubbles under the cover slip, place one edge of the cover into the lactophenol cotton blue first, and then lower the remaining part of cover slip gently until it is flat on the slide.

Mold lings to the cover slip. Examine your wet-mount under the low-power objective; then swing the high-power objective into place. Make drawings of all specialized structures of mold that you can find. Label the structures appropriately. Draw mold structures in your laboratory notebook and indicate all observed parts.

LABORATORY REPORT

Date:
Name:

Lab Title: Morphological Characteristics ofYeasts and Molds

1. **Draw a few representative yeast structures from microscopic analysis.**

   (a) Cultural characteristics of molds from PDB:

   (b) Morphology of molds on PDA:

2. **Draw a few representative mold structures from cultural characteristics.**

   (a) Cultural characteristics of yeast from PDB:

   (b) Morphology of yeasts on PDA:

3. **Draw a few representative structures from mold microculture.**

   (a) Cultural characteristics on PDA

   (b) Morphology of mold from direct microscopic analysis of culture:

   (c) Morphological characteristics from wet-mount preparation.

4. **Compare yeasts and molds in their macroscopic and microscopic structures.**
5. **Describe your observations of spores on the wet-mount slide.**
6. **Explain importance of morphological structures of molds.**
7. **What is advantage of slide culture over the plate culture?**

PRACTICE 25

Isolation, enumeration, and uses of bacteriophages

25.1 Theoretical information

There are different bacteriophage families in the different orders. Some of the orders and their families are as following. Families of bacteriophages in Caudovirales order are: Ackermannviridae, Myoviridae, Siphoviridae, and Podoviridae. Families of bacteriophages in Ligamenvirales order are: Lipothrixviridae and Rudiviridae. Families of bacteriophages in Unassigned order are: Ampullaviridae, Bicaudaviridae, Clavaviridae, Corticoviridae, Cystoviridae, Fuselloviridae, Globuloviridae, Guttaviridae, Inoviridae, Leviviridae, Microviridae, Plasmaviridae, Pleolipoviridae, Portogloboviridae, Sphaerolipoviridae, Spiraviridae, Tectiviridae, Turriviridae, and Tristromaviridae. The families include many bacteriophages. They can be classified according to morphological structures. Some bacteriophage shapes are given in Fig. 25.1. Bacteriophages are grouped under six types according to their head, tail structure, and shapes.

Type A. This group is a hexagonal head plate at the base of the tail. It has a capsid that gives the shape to the head and consists of similar sub-forms. The tail has a hollow tube and has a sheat around that can contract like a spring. DNA is found in the middle of the head. Some of the bacteria such as T2, T4, and T6 phages of *Escherichia coli*, *Salmonella* Typhi, some other *Salmonella*, *Pseudomonas*, and *Staphylococcus* include these phages. Bacteriophages attach to this type of bacteria with their tails.

Type B. The structure of the head parts is like those of type A phages and the tails of these phages with double-stranded linear DNA are long. However, it does not have a contractile sheath around it. *E. coli* T1, T5, and λ phages are in this group.

Type C. Its distinction from type B is its short tail. This group contains a double-stranded linear DNA in the capsid. T3, T7 phages, some *Salmonella* phages, and some of *E. coli* phages have this feature.

Type D. It is a single-stranded circular DNA phage that has no tails and is icosahedral in appearance. Its capsomers of the capsid are large. *E. coli* QX 174 phage is an example of this type.

https://doi.org/10.1016/B978-0-323-91017-0.00020-2

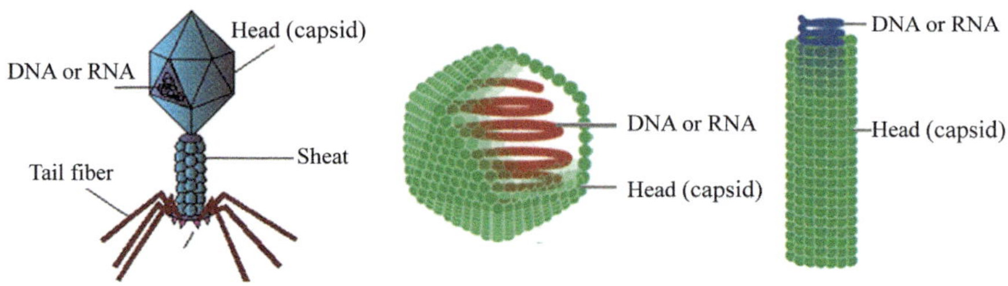

Fig. 25.1 Some schematic view of bacteriophages.

Type E. This type is also tailless phages. Apart from the small size of its capsomers, the appearance is like D icosahedral. Its core has a single-stranded linear RNA. Phages showing the specificity of *E. coli* adhesion to the sex pillar are of this nature. F2 R17, fr, and MS2 phages are other examples of this type.

Type F. This group has a completely different appearance and all are flamenco appearance. One of them, fd phage, is made of 800 nm length and 6 nm width, with a linear single-stranded DNA in the core and a protein sheath around it. fd phage is one of the *E. coli* phages.

25.2 Uses of bacteriophages

Bacteriophages cannot infect and replicate in human cells; they can conduct a critical mediator of genetic exchange between pathogenic and non-pathogenic bacteria. The transfer of genes from one bacterial strain to another by a bacteriophage is called transduction. Genetic exchange can alter microbial populations.

Bacteriophages can be used to identify certain strains of bacteria. By testing the susceptibility of bacteria to phages, it is possible to distinguish and type various bacterial strains. Various bacterial strains originate from the same source of the bacteriophage type. Taking advantage of this, bacteriophages are used in epidemiological studies. In this way, it is possible to determine whether the infections (bacterial or viral) originate from the same source and the ways of infection spread. The phage type determination of *Staphylococcus aureus* and some intestinal bacteria (such as *Salmonella*, *Shigella*, *E. coli*, and *Vibrio*) has been brought into standard phages by international committees.

Bacteriophages are clinically significant for several reasons. (i) Many pathogenic bacterial toxins are encoded by bacteriophage genomes. Bacteriophages can associate with a phage-borne toxin-mediated disease. Examples are cholera toxin in *Vibrio cholerae*, diphtheria toxin in *Corynebacterium diphtheriae*, botulinum neurotoxin in *Clostridium botulinum*, binary toxin in *Clostridium difficile*, and shiga toxin in *Shigella dyssanteriae*. Without their phage-encoded toxins, these bacterial species may either much less pathogenic or not pathogenic at all. (ii) Many microbial resistance mechanisms to antibiotics are provided by phage. Bacteriophages are vectors, which may introduce antimicrobial resistance genes into bacteria. (iii) Clinical use of lytic bacteriophages as an antimicrobial therapy (phage therapy) is possible in humans. Bacteriophages may be used as an alternative to antibiotics. They are a possible therapy against multi-drug-resistant strains of many bacteria. (iv) Phages are much more immunogenic than antimicrobial drugs and are rapidly cleared from the blood by the reticular endothelial system. (v) Bacteriophages can be used as a biological indicator of the presence of pathogenic bacteria in a complex environment, in the water or in the food. The biological indicator can be widely used in the detection of fecal contamination of water resources. If the indicator phage can multiply in the sample taken from this environment, it indicates that the pathogenic bacterium present in the source of the sample taken. Coliphages can be used as an indicator for water since coliphages can remain in the aquatic environment for a long time and show resistance to treatment methods. They can be detected easily, quickly, and cost-effectively in waste water and fecal residues. They can be used as a microbiological quality indicator for pool and surface waters. They are bacteriophages specific to *E. coli*. Water can be stored at 4°C at least 48 h without any significant change in the number of bacteriophages and small volume samples can be kept for months when stored at −20°C or stored at −80°C after adding 10% (v/v) glycerol.

Bacteriophages are used in the detection of bacteria because of their ability to distinguish strains of the same bacterial species. Many types of bacteria are sensitive to one type of bacteriophage, as well as more than one phage. While some strains in the bacterial species may resist to some phages, they may be sensitive to other phages. Each bacterial strain can be systematically infected with a standard phage panel to determine the sensitivity or resistance of each bacterial strain to each phage type. For example, in phage typing of *S. aureus* strains, it can be performed using an internationally standardized bacteriophage panel. Prior to the development of molecular methods, phage typing is used to monitor pathogenic bacterial strains for epidemiological purposes.

Bacteriophages are not as easily inactivated as vegetative bacterial cells, but they are vulnerable to UV inactivation and standard disinfection procedures. Lysogenic bacteriophage genetic material can be induced to switch to the lytic replication cycle by host cell stress, including antimicrobial drugs. Therefore, antibiotic treatments, especially those targeting gut microbial flora, can be expected to alter the viral (phage) microbiome as well.

25.3 Life cycle of bacteriophages

Bacteriophages are found wherever bacteria exist. There are more than 10^{31} bacteriophages on Earth. The main natural sources for bacteriophages and other viruses are seawater and sewage water. Phages can also appear in biofilms. Like all viruses, bacteriophages are very species specific regarding their hosts and usually infect a single bacterial species. Bacteriophages replicate within the bacterial cell following the injection of their genome into bacterial cytoplasm (Fig. 25.2). Once a bacteriophage attaches to a susceptible bacterial cell, it injects own genetic material (or completely enter), it enters one of the life cycles in the bacterial cell: lytic or lysogenic.

In lytic cycle (Fig. 25.2), a phage attaches to a susceptible bacterium, introduces its gene into the bacterial cell cytoplasm, and controls the ribosomes of the cell to synthesize phage proteins. The bacterial cell resources are rapidly converted to viral genomes and capsid proteins formation, which assemble into multiple copies of the original bacteriophage. Lysis of bacterial cell is achieved by an enzyme called endolysin, which attacks and breaks down the cell wall peptidoglycan. As soon as the cell is destroyed, the new phages can find new bacterial cells to infect. Some phages will not immediately lyse out of the cell if extracellular phage concentrations are high.

In the lysogenic replication cycle, the phage also attaches to a susceptible bacterium and introduces its gene into the bacterial cell cytoplasm. The bacteriophage gene is integrated into the bacterial cell chromosome or maintained as an episomal element (plasmid) relatively harmlessly in cytoplasm. In both cases, it is replicated and passed to daughter bacterial cells without killing them. Integrated phage gene is termed prophage, and the bacterial cell containing phage gene in own DNA is termed lysogen lysogenic bacterium. The prophage, which binds to the genetic structure in the bacterial cell, may return to the lytic replication cycle at any time, mostly under the influence of environmental conditions (such as presence of antibiotics). For example, the phage lambda of *E. coli* first enters the lysogenic cycle and then can enter the lytic cycle. Prophages are not always harmful to the bacterial cell. Sometimes prophage can add new function to the bacterial genome as

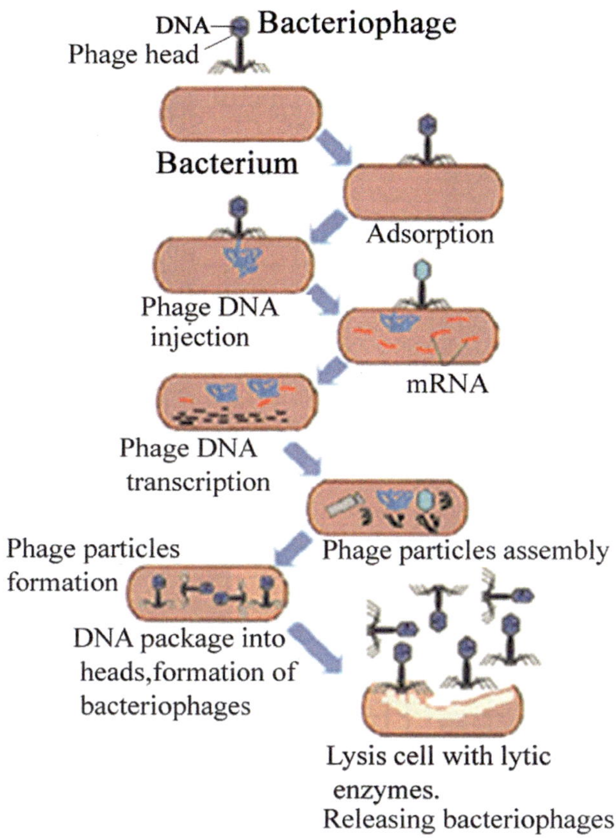

Fig. 25.2 Bacteriophage lytic cycle.

a result of lysogenic transformation, which may give bacteria a new feature or benefit. Examples to the bacteria can be transformed into highly virulent pathogenic strains from harmless strains of *Corynebacterium diphtheriae* or *Vibrio cholerae* with prophages, causing diphtheria or cholera respectively.

In solid media, the phages multiply on the bacteria and when they lysis them, they form lytic (clear) areas on the solid medium. These areas of lysis are called plaques. In areas where single phage falls with appropriate dilutions, each plaque represents a phage. In liquid culture, phages spread within the medium. Bacteria must be in the log growth stage to obtain optimal plaque formation, which is the growth indicator of bacteriophages. For this, the bacterial culture is added to the new liquid medium and incubated in a shaking water bath for 2–3 h at an optimum growth temperature until visible turbidity.

25.4 Bacteriophage culturing techniques

Materials

Microbial cultures: 24 h brain heard infusion broth (BHIB) culture of *E. coli* and *Staphylacoccus aureus*.

Solutions and chemicals: BHIB (10 mL in tube), 100 mL BHIB (in 250 mL flask), brain heard infusion agar (BHIA, in 150 mm Petri dish in diameter), 5 mL soft BHIA stab agar (0.7% agar), BHIA plates (15 mL per plate), and decastrength phage broth (DSPB) in metal-capped tube. $CaCl_2$ (400 μg/mL) must be added into all media before sterilization.

Equipments: Water bath (at 45°C), sterile Petri dishes (150 mm in diameter), sterile Erlenmeyer flask (250 mL), graduate, centrifuge, centrifuge tubes, sterile membrane filter assemblies (funnel), glass spreader (L-shaped), clamp, vacuum pomp, vacuum flask, sterile membrane filters (0.4 μm pore size), sterile Erlenmeyer flask, forceps, sterile 5 and 1 mL pipettes in box, sterile Tris-buffer water (in 150 mL flask) and flask of sewage water.

25.4.1 Identification of bacteriophages

Lower number of bacteriophages may be present in the sample. Therefore, the number of bacteriophages must be increased in sample using enriched medium. Enriching increases number of bacteriophages and they are easily isolated from this medium.

Enriching bacteriophages. Numbers of bacteriophages can be increased in enrichment medium from sewage water. For this aim, add 5 mL DSPB and 5 mL *E. coli* specific to coliphage into 45 mL of sewage water. The DSPB is incubated at 37°C for 20 h.

1. Inoculate 1 mL of *E. coli* culture into BHIB and incubate at 37°C until visible turbidity.
2. After incubation of enrich broth, dispense the enriched broth containing sewage into centrifuge tubes and fill each tube up to ½. Place the tubes to the centrifuge and centrifuge the tubes at 2500 rpm for 10 min.
3. Take liquid and add 10 mL Tris-buffer water and mix. Repeat centrifugation two times by adding Tris-buffer water into filtrate and mix gently.
4. Finally add 10 mL Tris-buffer water and mix gently. Aseptically filter through sterile filter paper (pore size 0.4 μm) to separate phages from *E. coli*.
5. Aseptically transfer the filtrated liquid from vacuum flask to a sterile Erlenmeyer flask closed with cotton plug. Be sure to flame the necks of flasks through work.

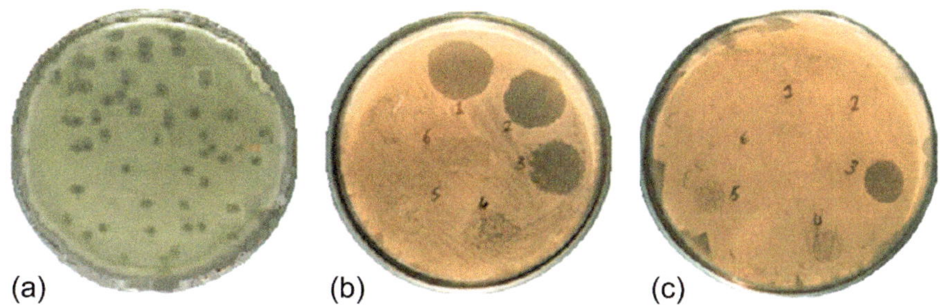

Fig. 25.3 (a) Diluted bacteriophage plagues, (b) number 2 is the routine test dilution of bacteriophage and (c) bacteriophage typing (number 3 indicates specificity of bacteriophage with the tested bacterium).

6. Transfer 1 mL of filtrate with 1 mL pipette into melted soft BHIA stab (5 mL in the tube at 45°C).
7. Transfer 0.01 mL of *E. coli* at visible turbidity with 1 mL pipette to filtrate added BHIA stab and mix gently.
8. After flaming the neck of the soft agar tube, pour the contents of soft agar onto BHIA plate.
9. After the agar is solidified, place the plates into 37°C incubator. If possible, examine the plates after 3 h to look for plaque formation (Fig. 25.3a). If some plaque is visible, measure them and record their diameters. Plaque size should be checked every hour for changes during 20 h incubation.
10. After incubation, clear area (zone) on BHIA in plates indicate presence of coliphage in sewage water.

25.4.2 Enumeration of bacteriophages from environmental sample

In this practice, coliphages (specific to *E. coli*) will be counted from sewage water to indicate quality of water.

1. Obtain a sewage water for bacteriophage analysis.
2. Dilute sewage water in 1:10 and 1:100 ratio using sterile Tris-buffer water. To do this dilution, transfer 1.0 mL of sewage water into 9 mL of Tris-buffered water and mix in vortex so that you prepare 1:10 dilution for sewage water. Then take 1.0 mL of 1:10 diluted sewage water and transfer to 9 mL Tris-buffered water and mix in vortex so that you dilute sewage water 1:100 times.
3. Melt the BHIA soft agar stab (5 mL) in tubes in a steam bath or autoclave.

4. Soak the melted agar in a water bath for 15 min to allow it to set the temperature to 45°C.
5. Inoculate 1 mL of *E. coli* culture into BHIB and incubate at 37°C until visible turbidity.
6. Add 1 mL *E. coli* culture at visible turbidity and 1 mL of undiluted/ diluted sewage water into melted soft agar (at 45°C).
7. Remove the tubes from the water bath, mix the medium with shaking the tube between hands for 5 s.
8. Using a paper towel, wipe off the water around the tube and pour the soft agar medium onto the BHIA plate (containing 15 mL medium) under aseptic conditions.
9. Quickly rotate the Petri plate to spread the spilled agar medium so that the agar covers the entire surface.
10. After the agar has solidified, incubate inverted Petri plates at 37°C for 20 h in an incubator. Moisture on the cover of the Petri plate should not drip on the agar surface and the virus will spread to the agar surface.
11. After incubation, remove the plates from the incubator and examine. In Petri plate, bacteria grow in the cloud and small open zone spots called plaques. Each of these plaques are areas where bacteria are killed by viruses, and each plaque represents a virus from the sample (Fig. 25.3a).
12. Count the number of plaques from plates containing 30 and 300 plaques.
13. Record major differences in the size or appearance of the plaques.
14. Calculate number of phages in the sewage water using the following formula:

$$\text{Plaque forming units (pfu) / mL} = (\text{Average number of plaque counts / (dxv)})$$

where, d = dilution, v = volume of diluted sewage water used.

25.4.3 Identification of bacteria by bacteriophage

Bacterial species/strain can be identified using specific bacteriophage. In this technique, bacteriophages specific to *S. aureus* strain can be used to identify unknown *S. aureus*. Identification of bacterium is performed at Routine Test Dilution (RTD) of bacteriophages.

25.4.3.1 Identification of bacteria with known bacteriophage

(a) Activation of lyophilized bacteriophages

1. Open vial containing lyophilized phage under aseptic conditions.
2. Add 0.5 mL of BHIB into opened vial.

3. Mix and allow for 10 min to rehydrate the phages.
4. Inoculate 1 mL of phage specific *S. aureus* strain into BHIB (100 mL) in a 250 mL flask and incubate until visible growth.
5. Add 1 mL of rehydrated phage into *S. aureus* culture and incubate at 37°C for 20 h in the incubator.
6. After incubation, clear liquid culture indicates formation of phages. Use this activated phage culture in the identification tests.

(b) Routine Test Dilution

RTD of activated phage culture should be detected.

1. Inoculate phage specific *S. aureus* strain into the BHIB (10 mL) in the tube and incubated at 37°C until visible turbidity.
2. Take BHIA plates, mark the Petri plate at bottom with a diameter of 1 cm area and label each area with a phage culture dilution.
3. Remove 1 mL of *S. aureus* culture at visible growth using sterile 1 mL pipette. Add this culture onto BHIA plate.
4. Take a glass spreader, soak into ethanol in a beaker, pass through flame, burn the alcohol, and allow to cool near the flame.
5. Spread plate culture using flamed glass spreader (after soaking into alcohol) and dry the surface area of agar medium.
6. Dilute activated phage culture in Tris-buffer water up to 10^{-5} and measure their turbidity of each undiluted/diluted culture using spectrophotometer.
7. Record these turbidities with respect to dilution rate.
8. Add 0.01 mL of undiluted/diluted phage culture to the marked area on the agar medium in plate and spread on 1 cm area.
9. Allow to dry the phage drops.
10. After drying, incubate the inverted plates at 37°C for 20 h in the incubator.
11. After incubation, analyze phage colonies (plaques) and indicates the dilution that forms a honeycomb-shaped phage plaque on plate (Fig. 25.3b as number "2" indicating RTD) and record this dilution as RTD for tested bacteriophages. Also record turbidity of this RTD.

(c) Identification of bacteria in liquid medium

1. Add 1 mL of 24 h unknown culture assumed as *S. aureus* into BHIB in duplicate and incubate at 37°C until visible growth.
2. Add 0.01 mL of bacteriophage from RTD into *S. aureus* culture at visible growth and gently mix.
3. Incubate the tubes at 37°C for 20 h in the incubator. Every hour, check the tube whether the liquid becomes clear.
4. After incubation, clearing in the tube indicates specificities between bacterium and bacteriophage. The unknown bacterium is *S. aureus*.

(d) Identification of bacteria on solid medium

1. Add 1 mL of 24 h unknown culture assumed as *S. aureus* into BHIB in tube and incubate at 37°C in the incubator until visible turbidity.
2. Spread plate 1 mL of unknown culture at visible turbidity onto BHIA plate using glass spreader and allow to dry the surface area of agar near flame by quarter opening cover of plate.
3. On the reverse side of the plates, mark with a diameter of 1 cm area for each bacteriophage name at RTD.
4. Add and spread 0.01 mL of suspended bacteriophage from RTD onto 1 cm labeled area on the plate.
5. Maximum 6 bacteriophages can be placed on Petri plate (150 mm in diameter). Allow to dry the phage drops on BHIA.
6. After drying, incubate the inverted plates at 37°C for 20 h in the incubator.
7. After incubation, look at plaque formation on the plates. Plaque formation indicates strain number corresponding to phage specific bacterium. Therefore, it indicates the strain number for unknown *S. aureus* (Fig. 25.3c). In this figure, phage number 3 indicates specificities bacteriophage with the tested *S. aureus*.

25.4.3.2 Identification of bacteria with own bacteriophage

This typing is done according to the fact that the bacteriophages carried by the lysogenic bacterium.

1. Inoculate lysogenic *S. aureus* strain into BHIB and incubate at 37°C in an incubator until visible turbidity.
2. On the reverse side of the BHIA plates, mark the name of tested bacterium with a diameter of 1 cm area.
3. Using 1 mL pipette, add 1 mL of unknown *S. aureus* culture onto labeled BHIA plates.
4. Spread plate 1.0 mL lysogenic *S. aureus* culture containing on the BHIA plates. Allow to dry surface area of agar medium.
5. Add 0.01 mL of unknown *S. aureus* culture to marked area on the plate and spread on one cm area.
6. Allow to dry the bacteriophage drops.
7. After drying, incubate the inverted plates at 37°C for 20 h in the incubator.
8. After incubation, look at for the formation of plaques on the plates. Plaque formation indicates host strains carrying the characteristics of the phage and hence the type of the isolated unknown bacterium is *S. aureus* strain (Fig. 25.3c).

LABORATORY REPORT

Date:
Name:

Exercise 25: Isolation, Enumeration, and Uses of Bacteriophages

1. **Differentiate between the followings:**

Lysis:

Lysogeny:

2. **Differentiate between the followings:**

Virulent phage:

Temperate phage:

3. **In what way can bacteriophage alter the genetic structure of a bacterium?**

4. **What would be the explanation for the presence of plaques on the plate?**

5. **Explain importance of bacteriophage presence in the environmental sample with respect to food and food production.**

6. **Explain importance of pathogen detection in sample by bacteriophages.**

7. **Explain significance of bacteriophage DNA insertion to bacterial DNA.**

8. **What are the factors that determine the size of plaque?**

9. **Why did you use young growing culture?**

10. **Why did the plaques vary in size and shape?**

APPENDIX

Preparation of culture media, stains, and solutions

A.1 Preparation of culture media

The preparations of media are arranged in alphabetical order.

1. Amie's Transport Medium	
Charcoal pharmaceutical	10.0 g
Sodium hydrogen phosphate	1.15 g
Sodium chloride	3.0 g
Potassium dihydrogen phosphate	0.2 g
Potassium chloride	0.2 g
Sodium thioglycolate	1.0 g
Calcium chloride	0.1 g
Magnesium chloride	0.1 g
Agar	15.0 g
Distilled water	1000 mL

Dissolve the ingredients in the water by steaming. Adjust pH to 7.2. Bring to boil. Make up the loss due to evaporation. Dispense into test tubes. Sterilize in the autoclave for 15 min at 121°C.

2. Azide Dextrose Broth	
Beef extract	4.5 g
Tryptose	15 0.0 g
Dextrose	7.5 g
Sodium chloride	7.5 g
Sodium azide	0.2 g
Agar	15.0 g
Distilled water	1000 mL

Dissolve the ingredients in the water by steaming. Adjust pH to 7.2. Bring to boil. Make up the loss due to evaporation. Dispense into test tubes. Sterilize in the autoclave for 15 min at 121°C.

3. Brain Heard Infusion Agar	
Calf brains infusion	200.0 g
Beef hearts infusion	250.0 g
Proteose peptone	10.0 g
Dextrose	2.0 g
Sodium chloride	5.0 g
Disodium phosphate	2.5 g
Agar	15.0 g
Distilled water	1000 mL

Dissolve the ingredients except agar in the water by steaming. Adjust pH to 7.4. Add agar, bring to boil, and continue to heat until the agar is dissolved. Make up the loss due to evaporation. Dispense into test tubes and sterilize in the autoclave for 15 min at 121°C. Cool the medium to 50°C and pour into sterile Petri plates in a thick layer.

4. Brain Heard Infusion Blood Agar	
Calf brains infusion	200.0 g
Beef hearts infusion	250.0 g
Protease peptone	10.0 g
Dextrose	2.0 g
Sodium chloride	5.0 g
Disodium phosphate	2.5 g
Agar	15.0 g
Distilled water	1000 mL

Dissolve the ingredients except agar in the water by steaming. Adjust pH to 7.4. Add agar, bring to boil, and continue to heat until the agar is dissolved. Make up the loss due to evaporation. Sterilize in the autoclave for 15 min at 121°C. Cool the medium to 50°C and add aseptically 5% v/v of sterile sheep blood and pour into sterile Petri plates in a thick layer.

5. Brain Heard Infusion Broth	
Calf brains infusion	200.0 g
Beef hearts infusion	250.0 g
Proteose peptone	10.0 g
Dextrose	2.0 g
Sodium chloride	5.0 g
Disodium phosphate	2.5 g
Distilled water	1000 mL

Dissolve the ingredients in the distilled water. Adjust the pH to 7.4. Bring to boil. Dispense into test tubes and sterilize in the autoclave for 15 min at 121°C.

6. Deca Strength Phage Broth	
Peptone	100 g
Yeast extract	50 g
NaCl	25 g
K_2HPO_4	80 g
Distilled water	1000 g

Dissolve the ingredients in the distilled water. Adjust the pH to 7.0. Heat broth to dissolve ingredients. Dispense into test tubes and sterilize in the autoclave for 15 min at 121°C.

7. Eosin Methylene Blue (EMB) Agar	
Peptone	10.0 g
Lactose	5.0 g
Sucrose	5.0 g
Dipotassium phosphate (K_2HPO_4)	2.0 g
Eosin Y	0.4 g
Methylene blue	0.065 g
Agar	15.0 g
Distilled water	1000 mL

Dissolve the ingredients except agar in water by steaming. Adjust the pH to 7.2. Add agar, bring to boil, and continue to heat until the agar is dissolved. Make up the loss due to evaporation. Sterilize in the autoclave for 15 min at 121°C. Final pH should be 7.2. Avoid overheating. Sterilization reduces the methylene blue, leaving the medium orange color. The normal purple color of the medium may be restored by gentle shaking. If the reduced medium is not shaken to oxidize the methylene blue, a dark zone beginning at the top and extending downward through the medium will gradually appear. The sterilized medium normally contains a flocculent precipitate which should not be removed. By cooling to 50°C and gently agitating the medium before pouring it into plates, this flocculation will be finally dispersed.

8. Fluid Thioglycolate Agar	
Yeast extract	5.0 g
Sodium chloride	2.5 g
Pancreatic digest of casein	15.0 g
Sodium thioglycolate	0.5 g
Glucose (anhydrous)	5.0 g
Resazurin	0.001
L-Cystine	0.5
Agar	0.75 g
Distilled water	1000 mL

Dissolve the ingredients except agar in the water by steaming. Adjust the pH to 7.1. Add agar, bring to boil, and continue to heat until the agar is dissolved. Make up the loss due to evaporation. Dispense into test tubes, filling them to at least two-thirds of tube capacity and sterilize in the autoclave for 15 min at 121°C. Cool to 25°C. After storage of prepared medium, if more than 30% of the medium demonstrates oxidation (color change of indicator), reheat medium once in a boiling water or steam bath to drive off absorbed oxygen.

9. Litmus Milk	
Skim milk powder	100.0 g
Litmus (or azolitmin)	1.0 g

Dissolve the litmus in 10 mL of the water. Add aid of a few drops of 1 N NaOH. Dissolve the skim milk in 900 mL of distilled water. Adjust pH to 6.5. Add the indicator solution and sufficient water to make 1000 mL. Mix thoroughly. Dispense into test tubes and sterilize in the autoclave for 15 min at 121°C.

10. MacConkey Agar	
Peptone	20.0 g
Lactose	10.0 g
Bile salts	5.0 g
Sodium chloride	5.0 g
Neutral red	0.075 g
Agar	15.0 g
Distilled water	1000 mL

Dissolve the ingredients except agar in the water by steaming. Adjust pH to 7.4. Add agar, bring to boil and continue to heat until the agar is dissolved. Make up the loss due to evaporation. Sterilize in the autoclave for 15 min at 121°C. Dry surface of the gel before inoculation. Cool the medium to 50°C and pour into sterile Petri plates in a thick layer.

11. Malt Extract Agar	
Malt extract	6.0 g
Maltose	1.8 g
Dextrose	6.0 g
Yeast extract	1.2 g
Agar	15.0 g
Distilled water	1000 mL

Dissolve the ingredients in the water by steaming. Bring to boil and continue to heat until agar is dissolved. Distribute as required and sterilize in the autoclave for 15 min at 121°C (15 lb. pressure). Make up the loss due to evaporation. In order to inhibit the growth of bacteria, antibiotics may be added as sterile solutions to the molten medium immediately before pouring the plates; or the medium may be acidified to pH 4.5 or 3.5. Acidification of medium may be achieved by adding aseptically sterile 10% lactic acid (or citric acid) solution immediately before pouring into sterile plates. The exact amount of acid to be added will depend upon the make or even the batch of the constituents used an amount of lactic for 100 mL of sterile liquid medium to adjust the pH. Mix well and pour into sterile plates. Allow to harden. Never heat medium after the addition of acid, as heating in the acid state will hydrolyze the agar, reducing its solidifying properties so that the resulting medium will be soft. Cool the medium to 50°C and pour into sterile Petri plates in a thick layer.

12. Malt Extract Broth	
Malt extract	6.0 g
Maltose	1.8 g
Dextrose	6.0 g
Yeast extract	1.2 g
Agar	15.0 g
Distilled water	1000 mL

Dissolve the ingredients in the water by steaming. Adjust pH to 4.5 or 3.5. Distribute in 5 mL amounts into test tubes, and sterilize in the autoclave for 15 min at 121°C.

13. Motility Test Medium	
Tryptose	10.0 g
Sodium chloride	5.0 g
Agar	5.0 g
Distilled water	1000 mL

Dissolve the ingredients except agar in the water by steaming. Adjust pH to 7.2. Add agar, bring to boil, and continue to heat until the agar is dissolved. Make up the loss due to evaporation. Dispense in 8 mL amount into tubes and sterilize in the autoclave for 15 min at 121°C. Allow the medium to cool quickly with tubes in an upright position. Place tubes in cold water for best results.

14. MR-VP Broth	
Peptone	7.0 g
Dipotassium phosphate	5.0 g
Dextrose (glucose)	5.0 g
Distilled water	1000 mL

Dissolve the ingredients in the water by steaming. Adjust pH to 6.9. Dispense in 10 mL amounts into test tubes and sterilize in the autoclave for 15 min at 121°C.

15. Mueller-Hinton Agar	
Beef extract	300.0 g
Casamino acid	17.5 g
Starch	1.5 g
Agar	17.0 g
Distilled water	1000 mL

Dissolve the ingredients except agar in the water by steaming. Adjust pH to 7.4. Add agar, bring to boil, and continue to heat until the agar is dissolved. Make up the loss due to evaporation. Sterilize in the autoclave for, 15 min at 121°C. Cool the medium to 50°C and pour into sterile Petri plates in a thick layer.

16. Nitrate Broth	
Yeast agar	3.0 g
Peptone	3.0 g
Sodium chloride	3.0 g
Distilled water	1000 mL

Dissolve the ingredients in the water by steaming. Adjust pH to 7.0. Bring the solution to boil. Make up the loss due to evaporation. Dispense in 5 mL amounts into test tubes. Sterilize in an autoclave for 15 min at 121°C.

17. Nutrient Agar	
Yeast agar	3.0 g
Peptone	5.0 g
Sodium chloride	3.0 g
Agar	15.0 g
Distilled water	1000 mL

Dissolve the ingredients except agar in the water by steaming. Adjust pH to 7.0. Add agar, bring to boil, and continue to heat until the agar is dissolved. Make up the loss due to evaporation. Sterilize in the autoclave for 15 min at 121°C. Cool the medium to 50°C and pour into sterile Petri plates in a thick layer.

18. Nutrient Broth	
Yeast extract	3.0 g
Peptone	5.0 g
Sodium chloride	3.0 g
Agar	15.0 g
Distilled water	1000 mL

Dissolve the ingredients in the water by steaming. Adjust pH to 6.8. Bring to boil. Dispense into test tubes and sterilize in the autoclave for 15 min at 121°C.

19. Nutrient Gelatin	
Yeast extract	3.0 g
Peptone	5.0 g
Gelatin	120.0 g
Agar	15.0 g
Distilled water	1000 mL

Dissolve the ingredients in the water by steaming. Adjust pH to 6.8. Bring to boil. Make up the loss due to evaporation. Dispense into test tubes. Sterilize in the autoclave for 15 min at 121°C.

20. Phenol Red Glucose Broth	
Yeast extract	3.0 g
Peptone	10.0 g
Glucose	5.0 g
Sodium chloride	5.0 g
Phenol red	0.018 g
Agar	15.0 g
Distilled water	1000 mL

Dissolve the ingredients in the water by steaming. Adjust pH to 7.4. Bring to boil. Dispense into test tubes that are containing Durham tube and sterilize in the autoclave for 15 min at 121°C.

21. Phenol Red Lactose Fermentation Broth	
Peptone	10.0 g
Lactose	5.0 g
Lactose	5.0 g
Sodium chloride	5.0 g
Phenol red	0.018 g
Agar	15.0 g
Distilled water	1000 mL

Dissolve the ingredients in the water by steaming. Adjust pH to 7.4. Dispense into fermentation tubes that are containing Durham tube and sterilize in the autoclave for 15 min at 121°C.

22. SS Agar	
Lab-Lemco powder	5.0 g
Peptone	5.0 g
Lactose	10 0.0 g
Bile Salts	8.5 g
Sodium citrate	10.0 g
Sodium thiosulfate	8.5 g
Ferric citrate	1.0 g
Brilliant green	0.00018 g
Neutral red	0.025
Agar	15.0 g
Distilled water	1000 mL

Dissolve the ingredients in the water by steaming. Adjust the pH to 7.0. Bring to the boil with frequent agitation. Do not autoclave. Cool to about 50°C, mix and pour into sterile petri dishes.

23. Sabouraud Dextrose Agar	
Neopeptone	10.0 g
Dextrose	40.0 g
Agar	15.0 g
Distilled water	1000 mL

Dissolve the ingredients except agar in the distilled water by steaming. Adjust pH to 5.6. Add agar, bring to boil, and continue to heat until the agar is dissolved. Make up the loss due to evaporation. Dispense into test tubes or retain in flask and sterilize in the autoclave for 15 min at 121°C. Avoid overheating which could result in a softer agar medium, especially those with a low pH. Cool the medium to 50°C and pour into sterile Petri plates in a thick layer.

24. Selenite Broth	
Peptone	4.0 g
Lactose	4.0 g
Disodium hydrogen phosphate	5.0 g
Sodium hydrogen selenite	5.0 g
Potassium dihydrogen phosphate	5.0 g
Distilled water	1000 mL

Dissolve the ingredients in the water by steaming. Adjust pH to 7.0. Add agar, bring to boil, and continue to heat until the agar is dissolved. Do not sterilize in the autoclave. Dispense into 5 mL amount into sterile test tubes.

25. SIM Medium	
Yeast extract	3.0 g
Peptone	10.0 g
Tryptone	10.0 g
Dextrose	1.0 g
Ferrous ammonium sulfate	0.2 g
Sodium thiosulfate	0.2 g
Agar	3.5 g

Dissolve ingredients in water by steaming. Adjust the solution to pH 7.3. Add agar and melt agar in distilled water by heating in a steamer, with frequent mixing. Sterilize at 121°C for 15 min.

25. Simmons Citrate Agar	
$NH_4H_2PO_4$	1.0 g
K_2HPO_4	1.0 g
Sodium chloride	5.0 g
Sodium citrate	2.0 g
$MgSO_47H_2O$	0.2 g
Bromothymol blue	0.08 g
Agar	15.0 g
Distilled water	1000 mL

Dissolve the ingredients except agar in the water by steaming. Adjust pH to 6.8. Add agar, bring to boil, and continue to heat until the agar is dissolved. Make up the loss due to evaporation. Dispense into test tubes and sterilize in the autoclave for 15 min at 121°C.

26. Starch Agar	
Yeast extract	3.0 g
Peptone	5.0 g
Soluble starch	0.2 g
Agar	15.0 g
Distilled water	1000 mL

Dissolve the ingredients except agar with the water by steaming. Adjust pH to 7.5. Add agar, bring to boil, and continue to heat until the agar is dissolved. Make up the loss due to evaporation. Dispense into test tubes and sterilize in the autoclave for 15 min at 121°C.

27. Sucrose Broth	
Sucrose	10.0 g
Yeast extract	3.0 g
Peptone	5.0 g
Agar	15.0 g
Distilled water	1000 mL

Dissolve ingredients in water by steaming. Adjust the solution to pH 6.9. Add agar and melt agar in distilled water by heating in a steamer, with frequent mixing. Distribute into test tubes as required and sterilize at 115°C for 20 min.

28. Sulfide Indole Motility Test Medium

Yeast extract	3.0 g
Peptone	10.0 g
Tryptone	10.0 g
Dextrose	1.0 g
L-Lysine hydrochloride	10.0 g
L-Cystine	0.2 g
Ferric ammonium citrate	0.2 g
Sodium citrate	2.0 g
Sodium chloride	5.0 g
Gelatin	80.0 g
Bromocresol purple	0.02 g
Agar	4.0 g
Distilled water	1000 mL

Dissolve the ingredients except agar in the water by steaming. Adjust pH to 7.3. Add agar, bring to boil, and continue to heat until the agar is dissolved. Make up the loss due to evaporation. Dispense in 5–7 mL amount into test tubes and sterilize in the autoclave for 15 min at 121°C.

29. Transport Media

Sodium chloride	3.0 g
Potassium chloride	0.2 g
Calcium chloride	0.1 g
Magnesium chloride	0.1 g
Disodium phosphate	1.15 g
Sodium thioglycollate	1.0 g
Charcoal	10.0 g
Agar	15.0 g
Distilled water	1000 mL

Dissolve the ingredients except agar in the water by steaming. Adjust pH to 6.8. Add agar, bring to boil, and continue to heat until the agar is dissolved. Make up the loss due to evaporation. Dispense in 5 mL amount into test tubes and sterilize in the autoclave for 15 min at 121°C.

30. Triple Sugar Iron Agar	
Beef extract	3.0 g
Yeast extract	3.0 g
Peptone	15.0 g
Proteose peptone	5.0 g
Sodium chloride	5.0 g
Lactose	10.0 g
Sucrose	10.0 g
Dextrose	1.0 g
Ferrous ammonium sulfate	0.2 g
Sodium thiosulfate	0.2 g
Phenol red	0.024 g
Agar	13.0 g
Distilled water	1000 mL

Dissolve the ingredients except agar in the water by steaming. Adjust pH to 7.3. Add agar, bring to boil, and continue to heat until the agar is dissolved. Make up the loss due to evaporation. Dispense into test tubes in 7 mL amount and sterilize in the autoclave for 15 min at 121°C.

31. Tryptone Broth (for the indole test)	
Tryptone	3.0 g
Distilled water	1000 mL

Dissolve the tryptone in the distilled water. Bring to boil. Filter the solution through paper. Dispense in 5 mL amounts into test tubes. Sterilize in an autoclave for 15 min at 121°C.

32. Urea Agar	
Peptone	1.0 g
Dextrose	1.0 g
Sodium chloride	5.0 g
Potassium phosphate	2.0 g
Urea	20.0 g
Phenol red	0.012 g
Agar	15.0 g
Distilled water	1000 mL

Dissolve the ingredients except agar in the water by steaming. Adjust the pH to 6.8. Add agar, bring to boil, and continue to heat until the agar is dissolved. Make up the loss due to evaporation. Dispense into test tubes in 5 mL amount and sterilize in the autoclave for 15 min at 121°C.

A.2 Preparation of stains

The stain preparations are arranged in alphabetical order.

1. Acetone Alcohol	
Alcohol	700 mL
Acetone	300 mL

Mix alcohol and acetone and store in refrigerator.

2. Albert's Diphtheria Stain	
Toluidine blue	0.15 g
Methyl green	0.20 g
Acetic acid	1.0 mL
Alcohol	2.0 mL

Dissolve the toluidine blue and the methyl green in 100 mL distilled water. Then add the acetic acid and alcohol and mix well.

3. Boric Acid 0.2 M Solution	
Boric acid	12.4 g

Dissolve the boric acid in sufficient distilled water to make 1000 mL.

4. Carbolfuchsin (for simple staining)	
Ziehl-Neelsen's carbolfuchsin	10 mL

Dissolve the alcohol in sufficient distilled water to make 150 mL.

5. Crystal Violet (vital stain) 1:2000 Solution	
Crystal violet	0.5 g

Dissolve the crystal violet in 100 mL distilled water.

6. Crystal Violet (for simple stain solution)	
Crystal violet	1.0 g

Dissolve the crystal violet in 100 mL distilled water.

7. Gram Staining Reagents	
Gram Crystal Violet Solution	
Crystal violet	0.5 g

Dissolve the crystal violet in sufficient water to make 100 mL.

Gram Iodine Solution	
Iodine	1 g
Potassium iodide	2.0 g

Dissolve the iodine and potassium iodide in sufficient water to make 300 mL.

Gram Decolorizer	
Acetone	250 mL
Isopropyl alcohol	750 mL

Mix acetone and alcohol.

Dissolve the safranin in the alcohol. Add 100 mL distilled water and filter through paper.

Gram Safranin Solution	
Safranin	0.25 g
Alcohol	10.0 mL

8. Lactophenol-Cotton Blue (For wet-mounting and staining of molds)	
Saturated cotton blue solution	10 mL
Lactophenol	10 mL

Mix lactophenol and cotton blue solution and complete to 100 mL with distilled water.

9. Lactophenol-Picric Acid (For wet-mounting and staining of molds)	
Lactic acid	100 mL
Phenol	100 g
Glycerol	200 mL
Picric acid solution (saturated aq.)	100 mL

Dissolve the phenol in the picric acid solution without heat, then add the lactic acid and the glycerol.

10. Lactophenol Solution (For wet molds)	
Phenol crystals	100 g
Lactic acid	100 mL
Glycerol	200 mL

Dissolve the phenol in 100 mL distilled water and add lactic acid and glycerol.

11. Leifson's Flagella Stain

Solution A	
Basic fuchsine	1.2 g
Ethanol (95%)	100 mL

Dissolve with frequent shaking. Store in a tightly stoppered bottle to prevent evaporation.

Solution B	
Tannic acid	3.0 g

Dissolve Tannic acid in 100 mL distilled water. Add phenol to a concentration of 0.2% to prevent microbial growth.

Solution C	
Sodium chloride	1.5 g

Dissolve NaCl in 100 mL distilled water. These solutions are stable at room temperature. To prepare the working solution, mix equal quantities of solutions A, B, and C and store in a tightly stoppered bottle in the refrigerator (stable for several weeks) or deep freezer (stable for months). If store deep frozen, the stain should be well shaken after thawing since the ethanol tends to separate from the water.

Counter Stain	
Methylene blue	1.0 g

Dissolve the methylene blue in the alcohol. Add 100 mL distilled water and filter the solution through paper.

12. Methylene Blue 1% Solution	
Methylene blue	1.0 g

Add methylene blue in small portions into 100 mL distilled water in a beaker on a mortar. Mix until the dye is completely dissolved in water.

13. Methyl Red Indicator Solution (For methyl red test)	
Methylene blue	0.1 g
Alcohol	250.0 mL
Distilled water	150 mL

Dissolve the methyl red in the alcohol. Add 150 mL distilled water and filter the solution through paper.

14. Rose Bengal	
Rose Bengal	1.0 g
Calcium chloride ($CaCl_22H_2O$)	0.01 g
Phenol	5.0 g

Dissolve the phenol in 100 mL water. Add the other ingredients and stir until dissolved. Filter the solution through paper.

15. Ryu stain (For flagella wet-mount staining)
Solution I: 10 mL of 5% aqueous solution of phenol, 2 g of tannic acid, and 10 mL of saturated aqueous solution of aluminum potassium sulfate-12 hydrate.
Solution II: Mix 12 g crystal violet in 100 mL of 95% ethanol.
Final stain: Mix 1part solution II with 10 parts solution I and then filtering the mixture through filter paper to remove coarse precipitate.

16. Safranin Staining Solution (For simple staining)	
Safranin	0.25 g
Alcohol	10.0 mL

Dissolve the safranin in the alcohol. Add 100 mL distilled water and filter the solution through paper.

17. Loeffler's Methylene Blue (For simple staining)	
Methylene blue	0.3 g
Alcohol	30.0 mL

Dissolve methylene blue in 100 mL distilled water and frequently shaking.

18. Spore Stain	
Malachite Green	
Malachite green	5.0 g

Dissolve the malachite green in 100 mL distilled water.

Safranin	
Safranin	5.0 g

Dissolve the safranin in 100 mL distilled water.

19. Ziehl-Neelsen Staining Reagents (for acid-fast staining)	
Carbolfuchsin	
Basic fuchsine	1.0 g
Ethanol (95%)	10.0 g
Phenol, 5% aqueous solution	100.0 mL

Dissolve the basic fuchsine in the alcohol. Dissolve the phenol in the distilled water. Mix the two solutions. Filter through coarse filter paper. Resultant solution should be removed from light and re-filtered as necessary to remove crystal formation.

Acid-alcohol (Decolorizing agent)	
Hydrochloric acid (37%)	30.0 mL

Dissolve hydrochloric acid in alcohol (95%) and complete to make 1000 mL.

Loeffler's Methylene Blue	
Methylene blue	0.3 g
Alcohol	30.0 mL
Distilled water	100 mL

Dissolve the methylene blue in the alcohol. Add the distilled water and filter the solution through paper.

A.3 Preparation of solutions

The solution preparations are arranged in alphabetical order.

1. Alcohol 70% (by weight)	
Alcohol	700 mL

Dissolve the alcohol in sufficient distilled water to make 1000 mL.

2. Aqueous Iodine	
Iodine	5.0 g

Dissolve iodine in sufficient water to make 100 mL.

3. Barret's Reagent (Voges-Proskauer test reagents)	
Reagent A (α-naphthol)	
α-naphthol	5.0 g

Dissolve the α-naphthol in sufficient alcohol to make 100 mL.

Reagent B (KOH)	
Potassium hydroxide (KOH)	40.0 g

Dissolve KOH in sufficient distilled water to make 100 mL.

4. Chromic Acid Cleaning Solution		
Potassium dichromate	63.0 g	100 g
Distilled water	35 mL	750 mL
Concentrated sulfuric acid	960 mL	250 mL

Add the potassium dichromate to the water in a 2-liter flask. Slowly and carefully add the acid. Chromic acid is extremely corrosive. *Use protective clothing and eye protection.*

5. Formaldehyde	
Formaldehyde	5.0 g

Dissolve the formaldehyde in sufficient water to make 100 mL.

6. Kovac's Reagent	
p-dimethyl-amino-benzaldehyde	5.0 g
Amyl or isoamyl or butyl alcohol	75.0 g
Hydrochloric acid	25 mL

Dissolve the p-dimethyl-amino-benzaldehyde in the alcohol, then slowly add the hydrochloric acid. Store in the refrigerator.

7. Mercuric Chloride	
Mercuric chloride	5.0 g

Dissolve the mercuric chloride in sufficient water to make 100 mL.

8. Nitrite Test Reagents	
Solution A (Sulfanilic acid)	
Sulfanilic acid	8.0 g
Sulfur acid (concentrate)	48 mL

Add the sulfuric acid to 500 mL of distilled water. Add the sulfanilic acid and sufficient distilled water to make 1000 mL.

Solution B (Dimethyl-α-naphthylamine)	
α-naphthylamine	5.0 g
Sulfuric acid (concentrate)	8.0 g

Add the sulfuric acid to 100 mL distilled water and mix well. Then add the α-naphthylamine and stir until dissolved.

9. Nessler's Reagent	
Potassium iodide	7.0 g
Mercuric iodide	10.0 g
Potassium hydroxide	10.0 g

Dissolve the potassium iodide and mercuric iodide in 40 mL of distilled water. Dissolve the potassium hydroxide in 50 mL of distilled water and allow to cool. Mix the two solutions and add distilled water to 100 mL. Allow the precipitate to settle, decant the clear supernatant liquid into a reagent bottle, and discard the precipitate.

10. Phenol	
Phenol	5.0 g

Dissolve the phenol in sufficient distilled water to make 100 mL.

11. Ringers Solution (Diluent and suspending liquid)	
Sodium chloride	2.25 g
Potassium chloride	0.105 g
Calcium chloride	0.12 g
Sodium hydrogen carbonate	0.05 g

Dissolve the salts in 250 mL water and complete to 1000 mL with distilled water. Distribute as required and sterilize by autoclaving at 121°C for 20 min.

12. Silver Nitrate	
Silver nitrate	5.0 g

Dissolve the silver nitrate in sufficient distilled water to make 100.

Bibliography

[1] G. Alaeddinoglu, S. Kocabiyik, General Microbiology Laboratory Manual, Middle East Technical University Publishing, Ankara, 1986.

[2] S. Alnaimat, S. AbuShattal, Laboratory Manual in General Microbiology. Al-Hussein Bin Talal University Publishing, Ma'an, Jordan, 2012. 47p.

[3] Anonymous, Online Microbiology Lab Manuel, Department of Microbiology, Toronto, Canada, 2009.

[4] Anonymous, Basic Practical Microbiology: A Manuel, Society for General Microbiology, London, 2012.

[5] Anonymous, White Paper: An Overview of Virus Quantification Techniques, Virocyt, WP-2013-1, Chicago, 2013, pp. 1–5.

[6] Anonymous, Microbiology Laboratory Techniques, 2015 (Published on 02/03/2015) https://clinicalgate.com/microbiological-laboratory-techniques/#s0025.

[7] Anonymous, Basic Practical Microbiology: A Manuel, Microbiology Society, London, 2016. https://www.stem.org.uk/rx34go.

[8] Anonymous, Leica Microscope Manual. Leica Microsystems, Schweiz, Heerbrugg, 2019. 39p.

[9] Anonymous, Microbiology 101 – Introduction to Microbes. Science Courses, 2003–2020, 2020. https://study.com/academy/course/microbiology-101-intro-to-microbiology.html.

[10] A. Baer, K. Kehn-Hall, Viral concentration determination through plaque assays: using traditional and novel overlay systems, J. Vis. Exp. 93 (2014), e52065.

[11] S. Balakrishnan, S. Duraisamy, Zothansanga, N. Senthilkumar, G. Gurusubramanian, Practical Microbiology – A Laboratory Manual, Panima Publishing Corporation, New Delhi, 2013.

[12] M. Balouirin, M. Sadiki, S.K. Ibnsouda, Methods for in vitro evaluating antimicrobial activity: a review, J. Pharm. Anal. 6 (2) (2016) 71–79.

[13] H.J. Benson, Microbiological Applications: Laboratory Manual in General Microbiology, eighth ed., McGraw Hill, New York, 2001.

[14] J.G. Cappuccino, N. Sherman, Microbiology: A Laboratory Manuel, fifth ed., Addison Wesley Longman, Inc., California, 1999. 471p.

[15] M.A. Croxen, R.J. Law, R. Scholz, K.M. Keeney, M. Wlodarska, B.B. Finla, Recent advances in understanding enteric pathogenic *Escherichia coli*, Clin. Microbiol. Rev. 26 (4) (2013) 822–880.

[16] S.M. Doore, J.R. Schrad, W.F. Dean, J.A. Dover, K.N. Parent, *Shigella* phages isolated during a dysentery outbreak reveal uncommon structures and broad species diversity, J. Virol. 92 (8) (2018), e02117.

[17] O. Erkmen, Investigation on Nosocomial Staphylococcal Infections and their Control, Ph.D. Thesis, Gaziantep University Institute of Health Sciences, Microbiology Department, Gaziantep, 1994.

[18] O. Erkmen, Bacteriophage identification of *Staphylacoccus aureus* strains isolated from patients, personnel, and environmental samples in Gaziantep Medical Faculty hospital, Microbiol. Bull. 34 (2000) 293–302.

[19] O. Erkmen, T.F. Bozoglu, Food Microbiology Principles into Practice, Vol. 1: Microorganisms Related to Foods, Foodborne Diseases and Food Spoilage, John Wiley and Sons, Ltd., Chichester, 2016. 458p.

[20] O. Erkmen, T.F. Bozoglu, Food Microbiology Principles into Practice, Volume 2: Microorganisms in Food Preservation and Processing, John Wiley and Sons, Ltd, Chichester, 2016. 446p.

[21] R.I. Freshney, Culture of Animal Cells: A Manual of Basic Technique and Specialized Applications, seventh ed., John Wiley and Sons, Inc., Chichester, 2010. 728p.

[22] B.R. Glick, J.J. Pasternak, C.L. Patten, Bergey's Manual of Systematic Bacteriology. Volume Two: The Proteobacteria (Part C), second ed., Springer, New York, 2010.

[23] E. Goldman, L.H. Green, Practical Handbook of Microbiology, second ed., CRC Press, New York, 2009. 853p.

[24] T.A.T. Gomes, W.P. Elias, I.C.A. Scaletsky, B.E.C. Guth, J.F. Rodrigues, R.M.F. Piazza, L.C.S. Ferreira, M.B. Martinez, Diarrheagenic *Escherichia coli*, Braz. J. Microbiol. 47S (2016) 3–30.

[25] J. Grainger, J. Hurst, D. Burdass, Basic Practical Microbiology: A Manual, The Society for General Microbiology, Marlborough, 2001.

[26] A. Gucela, Microbiology: Practical Guide (A), 2010. https://www.academia.edu/37645526/MICROBIOLOGY_PRACTICAL_GUIDE_A_2010_2_PRACTICAL_I.

[27] C. Harley, Laboratory Exercises in Microbiology, ninth ed., McGraw-Hill Education - Europe, London, 2013. 512p.

[28] G.E. Kaiser, Biol 230: Microbiology Lab Manual, The Community College of Baltimore Country, Catonsville Campus, UK, 2012.

[29] N.R. Krieg, W. Ludwig, W.B. Whitman, B.P. Hedlund, B.J. Paster, J.T. Staley, N. Ward, D. Brown, A Laboratory Exercises in Microbiology, sixth ed., McGraw-Hill, New York, 2010.

[30] A.M. Mahasneh, S.M. Bdour, Microbiology Laboratory Manual, second ed., Academics for Publishing and Distribution Co, Amman, 2008. 124p.

[31] E. Matuschek, D.F.J. Brown, G. Kahlmeter, Development of the EUCAST disk diffusion antimicrobial susceptibility testing method and its implementation in routine microbiology laboratories, Clin. Microbiol. Infect. 20 (2013) O255–O266.

[32] B.A. Mistry, M.R. D'Orsogna, T. Chou, The effects of statistical multiplicity of infection on virus quantification and infectivity assays, Biophys. J. 114 (12) (2018) 2974–2985.

[33] A.M. Motlagh, A.S. Bhattacharjee, R. Goel, Biofilm control with natural and genetically-modified phages, World J. Microbiol. Biotechnol. 32 (4) (2016) 67–76.

[34] P.C. Okhuysen, H.L. DuPont, Enteroaggregative *Escherichia coli* (EAEC): a cause of acute and persistent diarrhea of worldwide importance, J. Infect. Dis. 202 (4) (2010) 503–505.

[35] I.L. Pepper, C.P. Gerba, Environmental Microbiology: A Laboratory Manual, second ed., Academic Press, Amsterdam, 2005. 209p.

[36] J. Petersen, S. McLaughlin, Laboratory Exercises in Microbiology: Discovering the Unseen World Through Hands-on Investigation, CUNY Academic Works Publishing, New York, 2016. 195p.

[37] R.A. Pollack, L. Findlay, W. Mondschein, R.R. Modesto, Laboratory Exercises in Microbiology, fifth ed., John Wiley and Sons, Inc., Chichester, 2018. 320p.

[38] A.H. Regua-Mangia, T.A.T. Gomes, M.A.M. Vieira, K. Irino, L.M. Teixeira, Molecular typing and virulence of enteroaggregative *Escherichia coli* strains isolated from children with and without diarrhoea in Rio de Janeiro city, Brazil, J. Med. Microbiol. 58 (2009) 414–422.

[39] A.J. Salle, Laboratory Manual on Fundamental Principles of Bacteriology, seventh. ed., McGraw-Hill Book, Inc., New York, 1973.

[40] J. Saxena, M. Baunthiyal, I. Ravi, Laboratory Manual of Microbiology, Biochemistry and Molecular Biology, Saxena, Baunthiya and Ravi, New Delhi, 2012. 379p.

[41] D.A. Schofield, N.J. Sharp, C. Westwater, Phage-based platforms for the clinical detection of human bacterial pathogens, Bacteriophage 2 (2) (2012) 105–283.

[42] F. Sirmatel, F.Y. Zeyrek, O. Erkmen, Determination of resistance to antibiotics by microdilution method in *Staphylococcus* strains of hospital origin, ANKEM J. 18 (4) (2004) 200–204.
[43] E. Tahtaci, I. Balci, O. Erkmen, *Staphylococcus aureus* carrier and bacteriophage typing of isolated strains in individuals working in food processing and marketing in the Gaziantep region, Food Technol. 1 (9) (1996) 53–57.
[44] K.P. Talaro, B. Chess, Foundations in Microbiology: Basic Principles, tenth ed., McGraw-Hill Science, New York, 2018. 608p.
[45] R.P. Tiwari, G.S. Hoondal, R. Tewari, Laboratory Techniques in Microbiology and Biotechnology, Abhishek Publications, Delhi, India, 2009. 189p.
[46] J. Vandepitte, J. Verhaegen, K. Engbaek, P. Rohner, P. Piot, C. Heuck, Basic Laboratory Procedures in Clinical Bacteriology, second ed., World Health Organization, Geneva, 2003.
[47] N. Varghese, P.P. Joy, Microbiology Laboratory Manual, Kerala Agricultural University, Pineapple Research Station, Kerala, 2016. https://www.researchgate.net/publication/306018042.
[48] W.B. Whitman, Bergey's Manual of Systematics of Archaea and Bacteria, first ed., Bergey's Manual Trust, John Wiley and Sons, Ltd., Chichester, 2015.

Index

Note: Page numbers followed by *f* indicate figures and *t* indicate tables.

N

O

P

Q

R

S

T

U

V

W

X

Y

Z

9780323910170